SPRINGER PROCEEDINGS IN PHYSICS 90

Springer-Verlag Berlin Heidelberg GmbH

Physics and Astronomy ONLINE LIBRARY

http://www.springer.de/phys/

SPRINGER PROCEEDINGS IN PHYSICS

79 **Nonlinear Dynamics and Pattern Formation in Semiconductors and Devices**
Editor: F.-J. Niedernostheide

80 **Computer Simulation Studies in Condensed-Matter Physics VIII**
Editors: D.P. Landau, K.K. Mon, and H.-B. Schüttler

81 **Materials and Measurements in Molecular Electronics**
Editors: K. Kajimura and S. Kuroda

82 **Computer Simulation Studies in Condensed-Matter Physics IX**
Editors: D.P. Landau, K.K. Mon, and H.-B. Schüttler

83 **Computer Simulation Studiesin Condensed-Matter Physics X**
Editors: D.P. Landau, K.K. Mon, and H.-B. Schüttler

84 **Computer Simulation Studies in Condensed-Matter Physics XI**
Editors: D.P. Landau and H.-B. Schüttler

85 **Computer Simulation Studies in Condensed-Matter Physics XII**
Editors: D.P. Landau, S.P. Lewis, and H.-B. Schüttler

86 **Computer Simulation Studies in Condensed-Matter Physics XIII**
Editors: D.P. Landau, S.P. Lewis, and H.-B. Schüttler

87 **Proceedings of the 25th International Conference on the Physics of Semiconductors**
Editors: N. Miura and T. Ando

88 **Starburst Galaxies**
Near and Far Editors: L. Tacconi and D. Lutz

89 **Computer Simulation Studies in Condensed-Matter Physics XIV**
Editors: D.P. Landau, S.P. Lewis, and H.-B. Schüttler

90 **Computer Simulation Studies in Condensed-Matter Physics XV**
Editors: D.P. Landau, S.P. Lewis, and H.-B. Schüttler

Series homepage – http://www.springer.de/phys/books/ssms/

Volumes 50–78 are listed at the end of the book.

D.P. Landau S.P. Lewis H.-B. Schüttler
(Eds.)

Computer Simulation Studies in Condensed-Matter Physics XV

Proceedings of the Fifteenth Workshop
Athens, GA, USA, March 11–15, 2002

With 82 Figures

Springer

Professor David P. Landau, Ph.D.
Professor Steven P. Lewis, Ph.D.
Professor Heinz-Bernd Schüttler, Ph.D.
Center for Simulational Physics
The University of Georgia
Athens, GA 30602-2451, USA

ISSN 0930-8989

ISBN 978-3-642-62423-0   ISBN 978-3-642-55522-0 (eBook)
DOI 10.1007/978-3-642-55522-0

Library of Congress Cataloging-in-Publication Data applied for

Bibliographic information published by Die Deutsche Bibliothek
Die Deutsche Bibliothek lists this publication in the Deutsche Nationalbibliografie; detailed bibliographic data is available in the Internet at <http://dnb.ddb.de>.

http://www.springer.de

Originally published by Springer-Verlag Berlin Heidelberg New York in 2003
Softcover reprint of the hardcover 1st edition 2003

Typesetting: Data conversion by A. Duhm
Cover concept: eStudio Calamar Steinen
Cover production: *design & production* GmbH, Heidelberg

Printed on acid-free paper   57/3141/XO   5 4 3 2 1 0

Preface

Over fifteen years ago, because of the tremendous increase in the power and utility of computer simulations, The University of Georgia formed the first institutional unit devoted to the use of simulations in research and teaching: The Center for Simulational Physics. As the international simulations community expanded further, we sensed a need for a meeting place for both experienced simulators and neophytes to discuss new techniques and recent results in an environment which promoted lively discussion. As a consequence, the Center for Simulational Physics established an annual workshop on Recent Developments in Computer Simulation Studies in Condensed Matter Physics. This year's workshop was the fifteenth in this series, and the continued interest shown by the scientific community demonstrates quite clearly the useful purpose that these meetings have served. The latest workshop was held at The University of Georgia, March 11–15, 2002, and these proceedings provide a "status report" on a number of important topics. This volume is published with the goal of timely dissemination of the material to a wider audience.

We wish to offer a special thanks to IBM Corporation and to the National Science Foundation for partial support of this year's workshop.

This volume contains both invited papers and contributed presentations on problems in both classical and quantum condensed matter physics. We hope that each reader will benefit from specialized results as well as profit from exposure to new algorithms, methods of analysis, and conceptual developments.

Athens, GA, USA
June 2002

D.P. Landau
S.P. Lewis
H.-B. Schüttler

Contents

Part II Quantum Systems

Part III Materials Properties

Part IV Liquids and Colloids

Part V Methods and Algorithms

1 Computer Simulation Studies in Condensed Matter Physics: An Introduction

D.P. Landau, S.P. Lewis, and H.-B. Schüttler

Center for Simulational Physics, The University of Georgia,
Athens, GA 30602-2451, USA

Computer simulation studies in condensed matter physics play an ever increasing role in many areas of investigation. The "status report" which is contained in this volume is the result of presentations and discussion that took place during the 15th Annual Workshop at the Center for Simulational Physics. The texts of both longer, invited presentations as well as a number of contributed papers are included. The reader will find that the scope of simulational/computational studies is broad and that substantial potential for cross-fertilization of methods between different sub-fields exists.

Demonstrating the renewed vigor in the magnetism community, Part I of the volume opens with seven papers that treat problems of current interest in magnetism. The first three papers all deal with magnetization switching. First, Novotny describes the application of the method of Monte Carlo with absorbing Markov Chains to the study of lifetimes of low temperature metastable states produced by reversing an applied magnetic field. He finds that the validity of theory at low temperature depends strongly on whether or not the field is at a value close to where the field dependent prefactor has a discontinuity. Rikvold et al. report on the integration of the Landau–Lifshitz–Gilbert equation for single domain nanomagnets. They find, in particular, that pillar shaped nanomagnets reverse through formation and propagation of a domain wall. Then, Brown et al. use the projective-dynamics method to find the saddle point in the free energy and determine the value of the magnetization at that point. Next, Schulenburg and Richter study the field induced magnetization in a strongly frustrated lattice containing quantum spins. They find that a macroscopic jump occurs close to the completely magnetized state. Ito and Ozeki then report a Monte Carlo study of the non-equilibrium relaxation in the $\pm J$ Ising model on a simple cubic lattice. They find universal behavior of the static exponents for a range of interactions, but the dynamic exponent varies continuously. Ozeki and Ito describe a study of the non-equilibrium relaxation to fully frustrated XY-models. They confirm the presence of two transitions and find that the chiral transition is not in the Ising universality class. In the last contribution to this section, Ujfalussy et al. describe a first principles calculation of the magnetic structure of FeMn/Co bilayers. They find that the structure of the magnetic ground state of FeMn changes from a non-collinear 3Q structure to a 1Q structure due to the proximity of the Co.

Part II treats quantum systems. Hoffmann and Nielaba have applied the path integral Monte Carlo method to investigate structural properties, phase transitions and quantum effects of condensates in cylindrical pore nanostructures. They present results for the density profiles and phase diagrams of Ar and Ne in cylindrical pores and make a detailed comparison of their path integral results to classical Monte Carlo results. Furukawa and Motome describe a new Monte Carlo algorithm for a certain class of strongly correlated electron systems with strong coupling to thermodynamically fluctuating classical fields, based on an expansion in terms of Chebychev polynomials. They present results obtained with this method for a simplified double-exchange model. Troyer et al. discuss their highly flexible and adaptable implementation of exact diagonalization algorithms which they have used to study the quantum dimer model in the context of quantum computing. They present evidence that the quantum dimer liquid ground state exhibits properties required for the design of a topologically protected quantum bit. Next, Todo has used a parallel implementation of the continuous-imaginary-time version of the Evertz–Lana–Marcu loop Monte Carlo algorithm, combined with an improved gap estimator, to study the integer-spin antiferromagnetic Heisenberg chain, up to spin $S = 3$. He reports new results for the zero-temperature correlation lengths and Haldane gaps, obtained from accurate, high-statistics finite size extrapolations. Schmid et al. have investigated the melting of stripes in the two-dimensional hard-core boson Hubbard model, using a recent variant of the "worm algorithm." They find that the melting at low temperatures involves a first order transition and they also conjecture the existence of a nematic phase in this system. In the next paper Pao and Bickers describe a study of the Mott insulator transition in the two-dimensional Hubbard model by means of a novel approximate, rotationally invariant Stratonovich–Hubbard decoupling scheme. The Monte Carlo evaluation of their resulting effective action is free of minus sign problems and allows them to obtain detailed low-temperature results for the Mott–Hubbard gap and for the static susceptibilities both at and near the half-filled-band limit. To complete this part, Maier et al. give diagrammatic derivations of the recently developed dynamical cluster approximation and of the closely related molecular coherent potential approximation for simulational studies of strongly correlated electron systems. They present a comparison of the convergence properties of both methods with increasing cluster size and show that the former converges faster for long-range properties, while the latter shows faster convergence for the local properties of the system.

In Part III a series of papers discusses the use of simulations to determine materials properties. To begin this section, Shim et al. use a strain percolation model designed to capture the essence of plastic deformation of metals under external stress. They examine the critical behavior of the model and find that the critical state is independent of the initial strain profile, indicating self-organizing behavior. They also show that the model displays long transient behavior, suggesting the importance of finite-size effects. Next, Puzyrev and Faulkner report on molecular dynamics simulations using embedded-atom potentials to study the concentration dependence of static and thermal atomic displacements in copper-gold alloys. Lastly, Nurminen et al. test the applicability of several flavors of empirical Si and Ge potentials for the study of both bare and island-covered Si (001) surfaces at finite temperatures. They observe notable edge effects of small Ge islands on this surface and identify two different sources of this behavior.

In Part IV three papers, all invited, discuss liquids and colloids. First, Chelikowsky et al. describe a new approach for simulations of liquids, based on quantum forces obtained by an ab initio pseudopotential density functional method. Their approach is illustrated by results for the dynamical structure factor, electric conductivity and diffusion constants of Si, GaAs, and ZnTe. Denniston and Robbins then demonstrate how molecular dynamics simulations can be used to map out mesoscopic continuum models of binary fluids. They present results for the order parameter profile, the density, and the stress, and they show how the simulations can reveal unexpected boundary condition terms in miscible binary fluids. In the last paper in this part, Horbach reports on electrokinetic phenomena in charged colloidal suspensions, studied using the Lattice–Boltzmann approach for solving the Navier–Stokes equations. The method is applicable beyond the weakly charged sphere limit and, as a specific example, results for a charged sphere sedimentation problem are presented.

Part V contains a series of papers on methods and algorithms. Wang et al. present a new algorithm for conducting efficient Monte Carlo simulations of Potts models. This algorithm averages over a large number of configurations in a single Monte Carlo sweep by mapping the formation and merging of "spin" clusters onto a summation over a binary tree. Building on their recently proposed "probability-changing cluster" algorithm for studying critical points, Tomita and Okabe extend its applicability to model systems possessing a vector order parameter. To test and illustrate the method, they examine the Kosterlitz–Thouless transition for the two-dimensional XY model and for q-state clock models. Schulz et al. unite the Wang–Landau energy-space sampling method and the N-fold way in order to increase the efficiency of simulations compared to the original single-spin-flip method. The next two papers consider numerical algorithms for solving the time-dependent Maxwell equations in materials with spatially varying permittivity and permeability. In the first of these, Kole et al. present a family of unconditionally stable algorithms based on Suzuki-product representations of the time-evolution operator for the electromagnetic fields. In this approach, unlike more standard methods, the stability of the time evolution is rigorously independent of the spatial mesh fineness. In the next paper, De Raedt et al. develop a highly efficient one-step algorithm for solving the time evolution of the Maxwell equations. Their approach is based on a convergent truncation of a Chebyshev expansion of the time-evolution operator and improves the efficiency by more than one order of magnitude, compared to standard approaches. To close this section, Swendsen discusses shortcomings of the standard statistical-mechanical definition relating the entropy of a system of many particles to the logarithm of the accessible phase space. He proposes an alternative definition based on the logarithm of the probability distribution and illustrates its main advantages over the standard definition.

Part I

Magnetism

2 Low-Temperature Metastability of Ising Models: Prefactors, Divergences, and Discontinuities

M.A. Novotny

Department of Physics and Astronomy, and
Engineering Research Center, Mississippi State University
P.O. Box 5167, Mississippi State, MS 39762-5167, U.S.A.

Abstract. The metastable lifetime of the square-lattice and simple-cubic-lattice kinetic Ising models are studied in the low-temperature limit. The simulations are performed using Monte Carlo with Absorbing Markov Chain algorithms to simulate extremely long low-temperature lifetimes. The question being addressed is at what temperatures the mathematically rigorous low-temperature results become valid. It is shown that the answer depends partly on how close the system is to fields at which the prefactor for the metastable decay either has a discontinuity or diverges.

2.1 Introduction

In many areas of science escape over a saddle point is an important physical phenomenon [1]. In some cases results can be derived in the low-noise limit [2]. However, what is often missing is any idea of when the low-noise limit is applicable. This paper addresses the question of how low the noise must be before the metastable lifetime becomes dominated by the low-noise limit results.

Here the lifetime of the metastable state of the kinetic Ising model is studied. The square-lattice Ising model is studied using Glauber dynamics [3], as well as a dynamic derived from a coupling of a quantum spin $\frac{1}{2}$ system to a one-dimensional phonon heat bath [4,5]. The simple-cubic dynamic Ising model with a Glauber dynamic is also simulated in the low-temperature limit.

Although only specific cases are simulated, a reasonable hypothesis emerges as to the question of how low is a low enough temperature before the low-noise results are valid. The hypothesis is that it depends on how close the applied field is to a value where the low-temperature prefactor has a discontinuity or a divergence. As the field where the low-temperature prefactor has a discontinuity or divergence is approached, the low-temperature limit is seen at progressively lower temperatures.

2.2 Models and Methods

The Hamiltonian of a spin-$\frac{1}{2}$ system can be written as

$$\mathcal{H} = \mathcal{H}_{\text{sp}} + \mathcal{H}_{\text{bath}} \,.$$

The spin Hamiltonian is

$$\mathcal{H}_{\rm sp} = -J\sum_{\rm nn}\sigma_i^z\sigma_j^z - H_z\sum_i \sigma_i^z \,, \tag{2.1}$$

where J is the ferromagnetic ($J > 0$) nearest-neighbor interaction parameter due to the exchange coupling between spins, H is the applied external field, σ_j^z is the z-component of the Pauli spin operator at site j, the first sum is over all nearest-neighbor (nn) pairs (4 nn for the square lattice and 6 for the simple-cubic lattice), and the second sum is over all N spins.

To simulate the quantum system given by the Hamiltonian requires explicit knowledge of the heat bath to which the spin system is coupled. In general, simulating the complete quantum-mechanical system to obtain the time dependence of the spin degrees of freedom is unnecessary. With the given spin Hamiltonian, $\mathcal{H}_{\rm sp}$, the dynamic is determined by the generalized master equation [4–6]

$$\frac{{\rm d}\rho(t)_{m'm}}{{\rm d}t} = \frac{i}{\hbar}[\rho(t),\mathcal{H}_{\rm sp}]_{m'm} + \delta_{m'm}\sum_{n\neq m}\rho(t)_{nn}W_{mn} - \gamma_{m'm}\rho(t)_{m'm}\,,$$

$$\gamma_{m'm} = \frac{W_m + W_{m'}}{2}\,, \qquad W_m = \sum_{k\neq m} W_{km}\,, \tag{2.2}$$

where $\rho(t)$ is the time dependent density matrix of the spin system, m', n, k, and m denote the eigenstates of $\mathcal{H}_{\rm sp}$, $\rho(t)_{m'm} = \langle m'|\rho(t)|m\rangle$, and W_{km} is a transition rate from the m-th to the k-th eigenstate. For our spin Hamiltonian there are no off-diagonal components, and the generalized master equation becomes identical to the classical master equation of a classical spin $\frac{1}{2}$ Ising system [7] with Hamiltonian

$$\mathcal{H}_{\rm Ising} = -J\sum_{\rm nn}\sigma_i\sigma_j - H_z\sum_i \sigma_i \,, \tag{2.3}$$

where $\sigma_j = \pm 1$ is the classical Ising spin. The explicit dynamic for the system depends on the transition rates W_{km}.

Martin in 1977 [8] used the quantum Hamiltonian of (2.1). He made the assumptions that each spin was coupled to its own fermionic heat bath, and that the correlation times in the heat bath are much shorter than the times of interest in the spin system. He then integrated over all degrees of freedom of the heat bath. He found that with appropriate assumptions the dynamic for the classical Ising model consisted of randomly choosing a spin and flipping it with a probability given by

$$p_{\rm G,flip} = q_{\rm G}\frac{\exp(\beta E_{\rm old})}{\exp(\beta E_{\rm old}) + \exp(\beta E_{\rm new})}\,, \tag{2.4}$$

where $\beta = T^{-1}$ (with Boltzmann's constant set to unity), $E_{\rm new}$ is the energy of the configuration with the chosen spin flipped and $E_{\rm old}$ is the energy of the original spin configuration. Here q_G is an attempt frequency, related to the microscopic coupling between the heat bath and the spin Hamiltonian. Note that to insure that all probabilities are between zero and one requires that $0 < q_G \leq 1$. In this paper $q_G = 1$. This derived dynamic corresponds the Glauber dynamic [3] of randomly choosing a spin, randomly choosing a random number r uniformly distributed between zero and one, and flipping the chosen spin if $r \leq p_{G,\text{flip}}$.

Recently with K. Park a dynamic was derived for the Ising model by coupling it to a phonon bath [4,5]. Here we concentrate on a one-dimensional phonon bath, but results for phonon baths in other dimensions have also been obtained [4,5]. Again the assumption is made that the correlation times in the heat bath are much shorter than the the times of interest in the spin system, and then the integration over all degrees of freedom of the heat bath is performed. In this case the dynamic is given by randomly choosing a spin, and by using a flip probability given by

$$p_{P,\text{flip}} = q_P \left| \frac{(E_{\text{old}} - E_{\text{new}}) \exp(\beta E_{\text{old}})}{\exp(\beta E_{\text{old}}) - \exp(\beta E_{\text{new}})} \right| \tag{2.5}$$

for $E_{\text{old}} \neq E_{\text{new}}$, and zero if $E_{\text{old}} = E_{\text{new}}$. In (2.5) the attempt frequency q_P must be chosen so as to make all probabilities less than one. For the fields and temperatures simulated here, this is accomplished by choosing $q_P = 0.01$.

We are most interested in measuring the lifetime of a metastable state for the Ising model. We start with all spins up ($\sigma = +1$), and apply a static field of strength H directed opposite to the spins (directed downward). The total magnetization for the model is $M = \sum_i^N \sigma_i$, so the magnetization starts at $M = N$. We measure the time τ required for the magnetization to reach $M = 0$, since for this magnetization one has crossed the saddle point and is rapidly moving toward the equilibrium magnetization. The units for τ used are Monte Carlo steps (mcs), where one mcs corresponds to one attempted spin flip. The physical time is proportional to the time in units of Monte Carlo steps per spin. However, the low-temperature limit relevant here is the Single-Droplet regime [9] (where a single nucleating droplet causes escape from the metastable state). Consequently, we will use mcs as our units of τ. In the present work the measurement for τ is repeated for 10^3 escapes using different random number sequences, and the average lifetime $\langle\tau\rangle$ is calculated from these escapes. The average lifetime in the Single-Droplet regime has the form

$$\langle\tau\rangle = A \exp\left(\beta \Gamma\right) / qts, \tag{2.6}$$

where A is the prefactor and $\Gamma(H)$ is the energy of the nucleating droplet. Note that the attempt frequency q enters this equation in a natural way, so that changing q will not change A. Here q will stand for q_G for the Glauber dynamic or q_P for the phonon dynamic. The prefactor A is a function of T and H, and depends on the explicit dynamic of the system. In a given field, Γ

and the prefactor A at zero temperature can be obtained from the measured lifetimes using a linear fit to

$$T \ln (\langle \tau \rangle q) = T \ln(A) + \Gamma \, . \tag{2.7}$$

Another way of analyzing the data [10], if the low-temperature value of Γ is known, is to calculate an effective prefactor A_{eff} at any finite temperature,

$$A_{\text{eff}} = \langle \tau \rangle \, q \, \exp(-\beta \Gamma) \, . \tag{2.8}$$

In (2.8) calculating A_{eff} assumes that the value of Γ used is the zero-temperature limit of Γ, i.e. the value of Γ derived in the low-noise limit.

The dynamic of randomly choosing a spin and flipping it or not with the decision made by comparing a random number r with the flip probability p_{flip} is the physical dynamic. Note the time is updated whether or not the spin is flipped. This dynamic cannot be changed without changing the physics. It can be implemented in a straightforward way on serial computers. However, the average lifetimes can be extremely long at low temperatures, making simulation in the straightforward manner unfeasible. Although the dynamic cannot be changed, the dynamic can be implemented on the computer in a more efficient fashion. One way of doing this is to use a rejection-free technique, by which only moves that are successful are implemented, and the time to make such a successful move is added to the current time. For discrete models, such as the Ising model, this method is called the n-fold way, and was implemented in continuous time by Bortz, Kalos, and Lebowitz in 1975 [11]. The n-fold way method can also be implemented in discrete time [12]. The n-fold way method can lead to exponential speed-ups in the algorithm. Unfortunately, particularly at small field values, the n-fold way exponential speedup is not sufficient to allow low-temperature Ising simulations to progress in a reasonable amount of computer time. However, the discrete-time version of the n-fold way can be further accelerated by realizing that the n-fold way algorithm uses a 1×1 absorbing Markov matrix to decide what will be the next spin configuration and to decide the time to exit from the current spin configuration. This Monte Carlo with absorbing Markov chain method (the MCAMC method) can naturally be generalized to the case of using $s \times s$ absorbing Markov Chains [13–15]. The details of the MCAMC method as well as the projective dynamics method, both of which are used here, can be found in a recent review [16]. For the simulations here, we have used $s = 1$, 2, and 4 in the absorbing Markov chains in the MCAMC method. For any s the average lifetime, $\langle \tau \rangle$, and all other averages are the same as for the straightforward implementation of the dynamics. This is because by using the MCAMC algorithm the dynamic has not been changed, the dynamic has just been implemented on the computer in a more intelligent fashion.

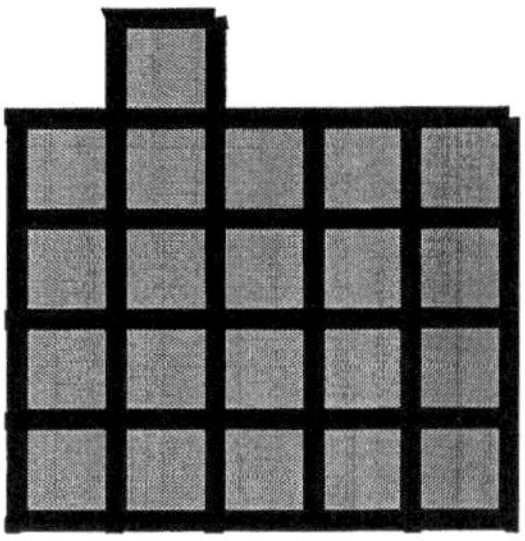

Fig. 2.1. The low-temperature square-lattice nucleating droplet with $\ell_2 = 5$ is shown

2.3 Square-Lattice Ising Model

At very low temperatures (in the low-noise limit) the kinetic Ising model lifetimes are influenced by the discreteness of the lattice. This has allowed for exact calculation of the saddle point as well as the most probable route to the saddle point. For the square lattice with $\ell_2 = \lfloor 2J/|H| \rfloor + 1$, the critical droplet is a square of size $\ell_2 \times \ell_2$ with one row removed and a single overturned spin on one of the longest sides [17]. Here $\lfloor x \rfloor$ denotes the integer part of x. Figure 2.1 shows one such critical droplet for $\ell_2 = 5$, when $\frac{2}{5}J < |H| < \frac{1}{2}J$. The average lifetime is then given by $\langle \tau \rangle q = A_2 \exp(\beta \Gamma_2)$ with $\Gamma_2 = 8J\ell_2 - 2|H|(\ell_2^2 - \ell_2 + 1)$. This is valid for low temperatures and for $|H| < 4J$ [17].

2.3.1 Glauber Dynamics

The prefactor for the square lattice with the Glauber dynamic was determined from absorbing Markov chain calculations to be $A_2 = \frac{5}{4}$ for $\ell_2 = 1$ and $A_2 = \frac{3}{8}$ for $\ell_2 = 2$ [14]. Recently the prefactor has been found to be $A_2 = 3/[8(\ell_2 - 1)]$ for $\ell_2 > 1$ [18]. At low enough temperature these results should hold, except when $2J/|H|$ is an integer. Figure 2.2 shows these prefactors for many values of H. Note that there is a discontinuity in the prefactor (but not in Γ_2) when $2J/|H|$ is an integer.

The exact values of Γ_2 and A_2 are known, so the MCAMC data at finite temperatures can be compared with these predictions. Figure 2.3a shows that near $|H| = 2J$ the predicted lines of $T \ln(\langle \tau \rangle)$ vs. T cross. If the measured lifetimes were to follow these expected curves, it would mean that for $T > 0.07J$ the lifetime at a field of $|H| = 1.99J$ would be smaller than for a field of $|H| = 2.01J$. In other words, there would be regions where the lifetime decreases as the field decreases. As seen in Fig. 2.3 this does not actually occur. Rather, the low-temperature predictions only agree with the data at lower and lower temperatures as the value of $|H|$ approaches the value $2J/|H|$ where the prefactor has a discontinuity. This is also demonstrated in Fig. 2.3b. Note that these MCAMC results using a Glauber dynamic are for 10^3 escapes from the metastable state for a 24×24 square lattice Ising system. The error

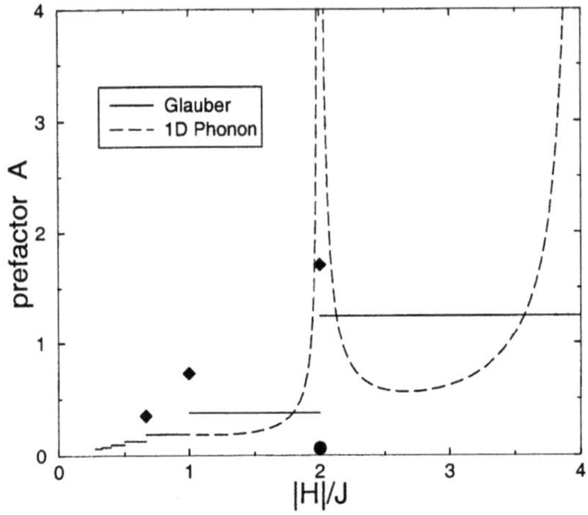

Fig. 2.2. The low-temperature square-lattice prefactors are shown vs. $|H|$. The solid lines and diamonds are the exact low-temperature results for the prefactors for the Glauber dynamic. The dashed curve and filled circle are the low-temperature prefactors for the Ising model with each spin connected to a $d = 1$ phonon bath. The curves and lines are derived analytically, while the symbols are from low-temperature MCAMC simulations. The prefactors are shown only for large values of $|H|$. Only three prefactors for the Glauber dynamic have been obtained at the discontinuity. Only prefactors for $|H| \geq J$ are shown for the phonon dynamic

estimates are from the second moment of τ obtained from the 10^3 escapes. These error estimates should be viewed as approximate, since the measured second moment most likely deviates substantially from the exact one when only 10^3 escapes are used.

Figure 2.4 shows the results for A_{eff} obtained at three different temperatures. Only MCAMC data points in the Single Droplet regime (as defined here by the ratio of the second moment of τ to $\langle\tau\rangle$ being greater than $\frac{1}{2}$ [9]) are shown. It is seen that the exact low-temperature prefactor is approached at finite temperature more quickly for values of H that are far from the values where A has a discontinuity. Note that in Fig. 2.4 the higher $|H|$ values start a cross-over toward the multi-droplet regime [9], and so should not be used in the comparison with the low-temperature prefactors. This cross-over depends on the system size as well as the temperature [9]. Figure 2.4b shows a portion of the data in Fig. 2.4a for $\ell_2 = 2$. Again note that near a value of H where there is a discontinuity (here $|H| = J$ and $2J$) the effective prefactor at finite temperature approaches the exact low-temperature prefactor extremely slowly.

A recent preprint [10] has performed a similar analysis of the prefactors for the square-lattice Ising dynamic using a generalized Becker–Döring approach. The authors find qualitatively similar results to those shown in

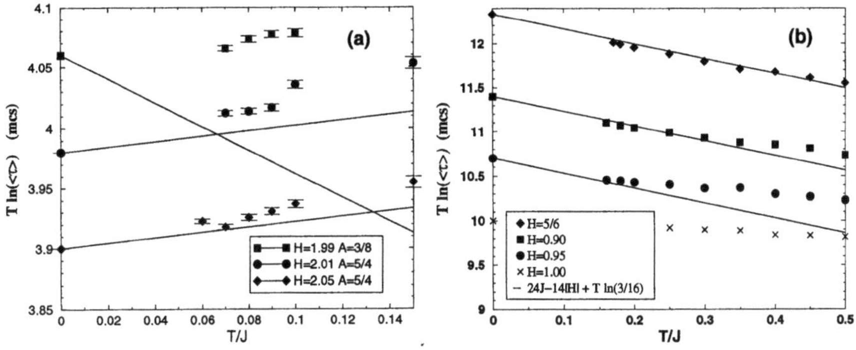

Fig. 2.3. The low-temperature results for the square-lattice Ising model as functions of T. The solid lines are the exact low-temperature results, including the prefactors. (**a**) Three different fields are shown. Note that the lines cross, which would mean a non-monotonic behavior of the lifetime in this region. (**b**) Results for $\ell_2 = 3$ are shown for several values of $|H|$. Note that as the discontinuity at $|H| = J$ is approached, the temperature required before the low-temperature results are seen decreases

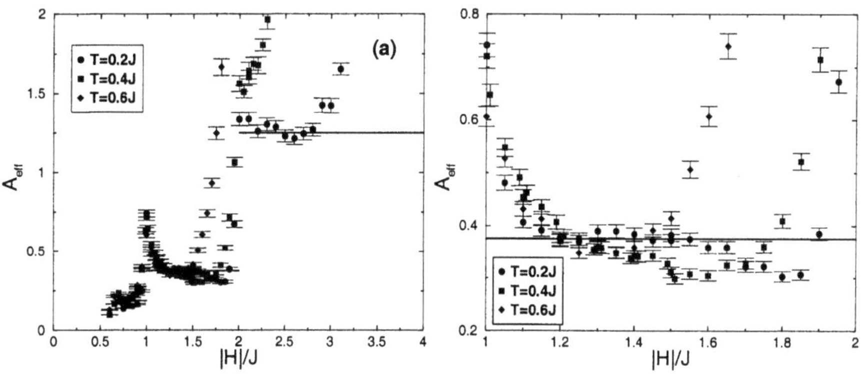

Fig. 2.4. The low-temperature results for the effective prefactor in the Single Droplet regime are shown as a function of $|H|$. The solid horizontal lines are the exact low-temperature prefactors. Three different temperatures are shown. Note that the deviation from the expected results at larger $|H|$ values can be mainly attributed to the cross-over between the Single and Multi-Droplet regimes. (**a**) For $\frac{1}{2}J \leq |H| \leq 4J$. (**b**) Only for $J \leq |H| \leq 2J$, where $\ell_2 = 2$

Fig. 2.4. However, unlike our results, their results do not approach the predicted low-temperature prefactors. They argue that this might be due to a difference between discrete and continuous times in the Glauber simulations. However, for the temperatures and field values simulated here we did not observe in our simulations that the prefactor value would depend on whether the time in the Glauber simulation is continuous or discrete. In the single-droplet regime relevant at low temperatures, the average lifetimes are extremely long, and consequently the average lifetimes should not depend in a significant way on whether the time in the simulation is continuous or discrete. The dynamic used in [10] is not identical to the Glauber dynamic, and this difference between the two dynamics might account for the difference in the prefactors.

2.3.2 Phonon Dynamics

For the square-lattice Ising model coupled to a one-dimensional phonon bath Fig. 2.2 shows the exact low-temperature prefactor [4] for $|H| > J$. This is given by the equation

$$A = \frac{4(2|H| - 4J) + (8J - 2|H|)}{4(2|H| - 4J)(8J - 2|H|)} \tag{2.9}$$

for $2J < |H| < 4J$ and

$$A = \frac{|H| + 2(2J - |H|)}{2^4|H|(2J - |H|)} \tag{2.10}$$

for $J < |H| < 2J$ [4]. The same nucleating droplets (except at $|H| = 2J$) are responsible for the decay of the metastable state for this dynamic as for the Glauber dynamic, so the same value of Γ_2 is found for both dynamics (except at $|H| = 2J$). Note that for this dynamic the prefactor is not constant when ℓ_2 is constant, and there is a divergence in the prefactor when $|H| = 2J$ or $|H| = 4J$. The divergence in the prefactor at $|H| = 2J$ is due to the nucleating droplet at this field being two next-nearest neighbor overturned spins that has $\Gamma = 8J$ rather than the value $\Gamma_2 = 4J$ for the Glauber dynamic. Recently, such a divergence in the prefactor has also been found in a model with continuous degrees of freedom [19]. Figure 2.5 shows values of A_{eff} (from finite-temperature MCAMC) at two different temperatures as they approach the exact low-temperature results. Note that the divergence in the prefactor is only seen at lower temperatures.

2.4 Simple-Cubic Lattice Ising Model

For the kinetic Ising model with the Glauber dynamic on a simple cubic lattice the average lifetime in the low-temperature limit (for fixed H and system size) is given by

$$\langle \tau \rangle = A_3 \exp(\beta \Gamma_3)/q$$

with

$$\Gamma_3 = 12J\ell_3^2 - 8J\ell_3 - 2|H|\ell_3^2(\ell_3 - 1) + \Gamma_2 \tag{2.11}$$

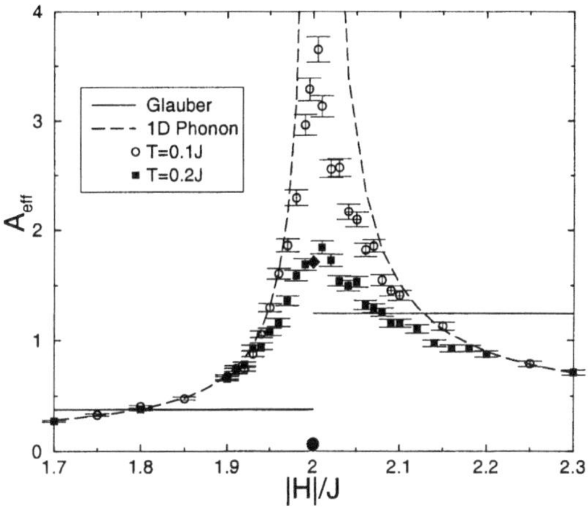

Fig. 2.5. The low-temperature results for the prefactors vs. $|H|$. The solid lines and diamond are the exact low-temperature results for the prefactors for the Glauber dynamic. The dashed curve and filled circle are the low-temperature prefactor for the Ising model with each spin connected to a $d = 1$ phonon bath. The curves and lines are derived analytically, while the symbols are from low-temperature Monte Carlo simulations. The effective prefactor from MCAMC data near $H = 2J$ is shown for 10^3 escapes from the metastable state on a 32×32 lattice at $T = 0.1J$ and $T = 0.2J$. Note that the low-temperature results are approached very slowly near the discontinuity

with $\ell_3 = \lfloor 4J/|H| \rfloor + 1$ [18,20,21] and Γ_2 the square-lattice nucleating droplet energy defined above. Equation (2.11) should be valid when $|H| \leq 4J$ and $2J/|H|$ is not an integer. The number of spins in the nucleating droplet is $n_c = \ell_3^2(\ell_3 - 1) + (\ell_2^2 - \ell_2 + 1)$. The nucleating droplet is a cube with ℓ_3 overturned spins per side, with one face removed and the appropriate square-lattice nucleating droplet placed where the face was removed. Figure 2.6 shows a nucleating droplet for $J \leq |H| \leq 2J$, so $\ell_3 = 3$ and $\ell_2 = 2$, and the nucleating droplet has $n_c = 21$. For this value of H the square-lattice nucleating droplet consists of three overturned spins in an L-shape. The prefactor for the simple cubic lattice with the Glauber dynamic has recently been found to be $A_3 = [16(\ell_3 - \ell_2 + 1)(\ell_2 - 1)]^{-1}$ for $\ell_3 \geq 3$ [18]. The low-temperature prefactor for $\ell_3 = 2$ (so $2J < |H| < 4J$) has not yet been derived, being a straightforward but lengthy calculation. The low-temperature prefactor for $\ell_3 = 1$ (so $4J < |H| < 6J$) is easily derived to be $A_3 = 7/6$.

Figure 2.7a shows the expected low-temperature result near $|H| = 2J$ for the simple cubic Ising model as a function of $|H|$. Projective dynamics [16,22,23] simulation results are also shown for $T = 0.2T_c$. The critical temperature is $T_c \approx 4.51J$. If the prefactor were unity, the expected low-temperature result (solid line) would be independent of temperature. For other prefactors, the dashed line is adjusted so the result with the prefactor

Fig. 2.6. A low-temperature nucleating droplet with $\ell_3 = 3$ and $\ell_2 = 2$. It consists of a cube of overturned spins with sides of length ℓ_3, with the spins from one face removed and a nucleating droplet with ℓ_2 placed on the removed face

depends on T. The simulation data are for a 32^3 Ising system using a projective dynamic method [16,22]. Note that away from a discontinuity the simulation data fall quite nicely on the expected low-temperature results including the predicted prefactor. However, they do not follow the expected results near the discontinuities (where ℓ_3 changes). Shown here is only the discontinuity near $|H| = 2J$. This is similar to the result found in the square-lattice model. Namely, how low in temperature the simulation must be performed before the low-temperature result is valid depends strongly on how close one is to a discontinuity in the low-temperature prefactor.

Note that, as seen in (2.11) and in Fig. 2.7a, there is a discontinuity in Γ_3 whenever ℓ_3 changes. For all values of H where ℓ_3 changes the discontinuity is such that Γ_3 decreases by $4J$. If the data were to follow these low-temperature results, as $|H|$ decreases through one of these discontinuities the average lifetime would decrease. This would occur not just because of a discontinuity in the prefactor, as in the square-lattice Ising model, but also because of a discontinuity in the exponential (in Γ_3).

Figure 2.8 shows MCAMC simulation results using the Glauber dynamic for the simple-cubic Ising model at $|H| = 3.5J$. For this value of H one has $\ell_3 = 2$ and $\ell_2 = 1$ so the nucleating droplet is predicted to have $n_c = 5$ overturned spins, and the predicted value of $\Gamma_3 = 5J$. The low-temperature prefactor is not known for this value of H. As seen in Fig. 2.8, the low-temperature MCAMC data do not tend toward $\Gamma_3 = 5J$, but could reasonably tend toward $\Gamma_3 = 7J$. The published low-temperature predictions [18,20,21] *do not agree* with our MCAMC results. The reason for this disagreement is that the previously reported low-temperature predictions are incomplete: they give accurate values for Γ_3 from (2.11) for some but not all values of H. This is because the most probable path of escape from the metastable state has a droplet with higher energy than predicted by Γ_3 from (2.11) for some val-

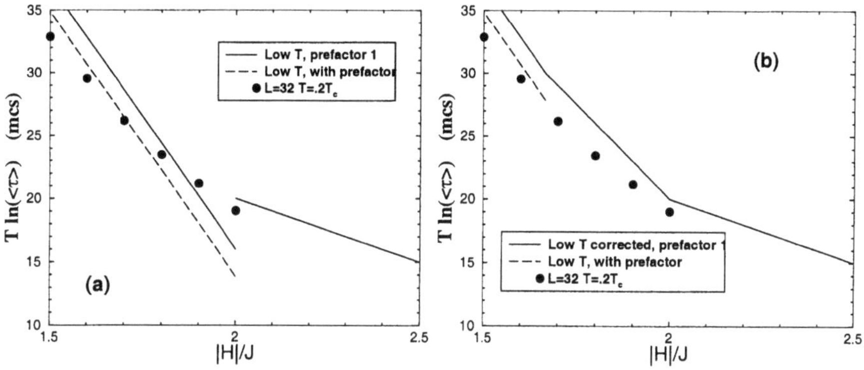

Fig. 2.7. The low-temperature results for the simple cubic lattice vs. $|H|$. The dashed line includes known exact prefactors. The symbols are projective dynamics simulation results on a 32^3 lattice at $T = 0.2T_c$. (**a**) The solid lines are the exact low-temperature results [18,20,21]. (**b**) The solid lines are the exact, corrected, low-temperature results. The dashed line shows the known exact prefactors, which in this figure are only known for $|H| < \frac{5}{3}J$. Note that now the exponential portion of the low-temperature prediction is continuous

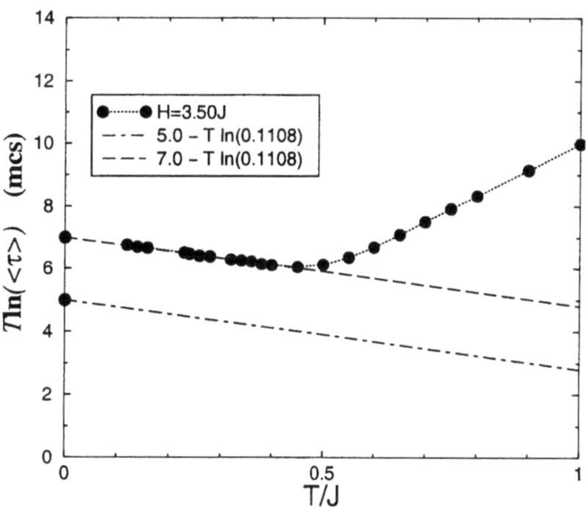

Fig. 2.8. The low-temperature results for the simple cubic lattice as a function of T for $|H| = 3.5J$. The lines have a slope of the best fit to the low-temperature data. The results at $T = 0$ are the exact low-temperature results, both the uncorrected ones (lower point) and the corrected ones (upper point)

ues of H. For $3J \leq |H| \leq 4J$ this higher energy saddle has an energy of $28J - 6|H|$, that corresponds to three overturned spins in an L-shape, and so has $n_c = 3$. For all $|H| < 4J$ this droplet lies along the most probable path to the saddle point, but it is the highest energy point along this path only for $3J \leq |H| \leq 4J$, and hence is then the critical nucleating droplet. This scenario predicts that for $|H| = 3.5J$ one should have $\Gamma_3 = 7J$, in agreement with the MCAMC results shown in Fig. 2.8.

Similar results hold near other values of H where ℓ_3 changes. For example, near $|H| = 2J$ the low-temperature predictions [18,20,21] are that for $n_c = 5$ $(2J < |H| < 4J)$ one has $\Gamma_3 = 40J - 10|H|$ and for $n_c = 21$ $(J < |H| < 2J)$ one has $\Gamma_3 = 100J - 42|H|$ (Fig. 2.6). However, there is a droplet with 15 overturned spins consisting of $2 \times 2 \times 3$ overturned spins with a $\ell_2 = 2$ droplet of overturned spins on one of the 2×3 faces. This 15-spin droplet has $\Gamma = 80J - 30|H|$. This is the nucleating droplet for values of H when this Γ is larger than the predicted values of Γ_3 for $n_c = 5$ or $n_c = 21$. This occurs for $\frac{5}{3}J < |H| < 2J$. At $|H| = 2J$ the droplets with 5 and with 15 overturned spins have the same energy, while at $|H| = \frac{5}{3}J$ the droplets with 15 and with 21 overturned spins have the same energy. The correct predicted values of $\Gamma_3(H)$ are hence continuous when this droplet with 15 overturned spins is included. This is shown in Fig. 2.7b. In general, one can show that all the predicted [18,20,21] discontinuities in Γ_3 where ℓ_3 changes vanish. After this study was completed another paper describing the low-temperature properties of the simple cubic dynamic Ising model [24] was brought to the author's attention. In [24] it is shown that the critical droplet size is $n_c = \ell_3(\ell_3 - \delta)(\ell_3 - 1) + (\ell_2^2 - \ell_2 + 1)$ where $\delta = 0$ unless $4J + \sqrt{16J^2 + |H|^2} < |H|(2\ell_3 - 1)$ in which case $\delta = 1$. For the simple cubic lattice, the critical value of $\Gamma_3(H)$ is thus continuous for all H, just as for the square lattice.

2.5 Summary and Conclusions

From long-time simulations of kinetic Ising models, the question of when the low-noise limit results are approached by finite-temperature simulations has been addressed. In particular, we have studied the limit where the system size and H are fixed, while the temperature is lowered. There are three main features that emerge:

- The square-lattice and simple-cubic-lattice low-temperature predictions have the form $\langle\tau\rangle = A\exp(\beta\Gamma)$ with $\Gamma(H)$ continuous almost everywhere, but with $A(H)$ discontinuous.
- The finite-temperature simulation results must be at a lower temperature to see the low-temperature predictions if the applied field is near a value where the low-temperature prefactor has a discontinuity or a divergence.
- The finite-temperature results have average lifetimes $\langle\tau\rangle$ which are decreasing functions of $|H|$. This is true even if the low-temperature results are not always decreasing as $|H|$ increases.

We have shown that these features are true for simulations in the square-lattice and simple-cubic-lattice kinetic Ising model with various dynamics. Whether or not these features of $\langle\tau\rangle$ can be shown to be true in general is an interesting question for future research.

Acknowledgements

Special thanks to M. Kolesik for allowing inclusion of unpublished projective dynamic data. Thanks to P.A. Rikvold and K. Park for many useful discussions. Partially funded by NSF DMR-0120310. Supercomputer time provided by the DOE through NERSC.

References

1. N.G. van Kampen: *Stochastic Processes in Physics and Chemistry* (Elsevier Science, Amsterdam 1982)
2. J.A. Bucklew: *Large Deviaion Techniques in Decision, Simulation, and Estimation* (Wiley, New York 1990)
3. R.J. Glauber: J. Math. Phys. **4**, 294 (1963)
4. K. Park, M.A. Novotny: Comp. Phys. Commun., in press; e-print cond-mat/0109214
5. K. Park, M.A. Novotny, in: *Computer Simulation Studies in Condensed Matter Physics XIV*. Ed. by D.P. Landau, S.P. Lewis, H.-B. Shüttler (Springer, Berlin Heidelberg New York 2002)
6. K. Blum: *Density Matrix Theory and Applications*, 2nd Edn. (Plenum Press, New York 1996) Chap. 8
7. K. Binder, E. Stoll: Phys. Rev. Lett. **31**, 47 (1973); K. Binder, H. Müller-Krumbhaar: Phys. Rev. B **9**, 2328 (1974)
8. P.A. Martin: J. Stat. Phys. **16**, 149 (1977)
9. P.A. Rikvold, H. Tomita, S. Miyashita, S.W. Sides: Phys. Rev. E **49**, 5080 (1994); P.A. Rikvold, B.M. Gorman, in: *Annual Reviews of Computational Physics I*, ed. by D. Stauffer (World Scientific, Singapore 1994) p. 149
10. V.A. Shneidman, G.M. Nita: e-print cond-mat/0201064
11. A.B. Bortz, M.H. Kalos, J.L. Lebowitz: J. Comput. Phys. **17**, 10 (1975)
12. M.A. Novotny: Comp. in Phys. **9**, 46 (1995)
13. M.A. Novotny: Phys. Rev. Lett. **74**, 1 (1995); erratum **75**, 1424 (1995)
14. M.A. Novotny, in: *Computer Simulation Studies in Condensed Matter Physics IX*. Ed. by D.P. Landau, K.K. Mon, H.-B. Shüttler (Springer, Berlin Heidelberg New York 1997) p. 182
15. M.A. Novotny: *Comp. Phys. Commun.*, in press; e-print cond-mat/0108429
16. M.A. Novotny, in: *Annual Reviews of Computational Physics IX*. Ed. by D. Stauffer (World Scientific, Singapore 2001) p. 153
17. E. Jordão Neves, R.H. Schonmann: Commun. Math. Phys. **137**, 209 (1991)
18. A. Bovier, F. Manzo: e-print cond-mat/0107376
19. R.S. Maier, D.L. Stein: Phys. Rev. Lett. **87** 022301 (2001)
20. D. Chen, J. Feng, M. Qian: Science in China (Series A) **40**, 832 (1997)
21. D. Chen, J. Feng, M. Qian: Science in China (Series A) **40**, 1129 (1997)
22. M. Kolesik, M.A. Novotny, P.A. Rikvold: Phys. Rev. Lett. **80**, 3384 (1998)
23. M.A. Novotny, M. Kolesik, P.A. Rikvold: Comp. Phys. Commun. **121–122**, 303 (1999)
24. G. Ben Arous, R. Cerf: Electronic J. Prob. **1**, 1 (1996)

3 Fine-Grid Simulations of Thermally Activated Switching in Nanoscale Magnets

P.A. Rikvold[1,2], G. Brown[1,3], and M.A. Novotny[4]

[1] School of Computational Science and Information Technology, Florida State University, Tallahassee, FL 32306-4120, USA
[2] Center for Materials Research and Technology and Department of Physics, Florida State University, Tallahassee, FL 32306-4351, USA
[3] Center for Computational Sciences, Oak Ridge National Laboratory, P.O.Box 2008 Mail Stop 6114, Oak Ridge, TN 37831-6114, USA
[4] Department of Physics and Astronomy, Mississippi State University, Mississippi State, MS 39762, USA

Abstract. Numerical integration of the Landau–Lifshitz–Gilbert equation with thermal fluctuations is used to study the dynamic response of single-domain nanomagnets to rapid changes in the applied magnetic field. The simulation can resolve magnetization patterns within nanomagnets and uses the Fast Multipole method to calculate dipole-dipole interactions efficiently. The thermal fluctuations play an essential part in the reversal process whenever the applied field is less than the zero-temperature coercive field. In this situation pillar-shaped nanomagnets are found to reverse through a local mode that involves the formation and propagation of a domain wall. Tapering the ends of the pillars to reduce pole-avoidance effects changes the energies involved but not the fundamental process. The statistical distribution of switching times is well described by the independent nucleation and subsequent growth of regions of reversed magnetization at both ends of the pillar.

Magnetic nanoparticles are important components of nanotechnology. Previously, Ising models, which describe only highly anisotropic materials, have been used to understand thermally activated switching in single-domain magnets [1,2]. Single-domain magnets with dimensions of only a few nanometers are being manufactured and measured, for instance, using scanning microscopy techniques [3]. We have chosen these nanomagnets as a specific system in which to investigate magnetization switching with more realistic computational models.

Here magnetic materials are modeled by position-dependent magnetization density vectors $\boldsymbol{M}(\boldsymbol{r})$ with fixed length M_s, that precess around the local field $\boldsymbol{H}(\boldsymbol{r})$ according to the Landau–Lifshitz–Gilbert equation [4,5]

$$\frac{\mathrm{d}\boldsymbol{M}(\boldsymbol{r})}{\mathrm{d}t} = \frac{\gamma_0}{1+\alpha^2}\boldsymbol{M}(\boldsymbol{r}) \times \left[\boldsymbol{H}(\boldsymbol{r}) - \frac{\alpha}{M_s}\boldsymbol{M}(\boldsymbol{r}) \times \boldsymbol{H}(\boldsymbol{r})\right] , \tag{3.1}$$

where $\gamma_0 = 1.76 \times 10^7\,\mathrm{Hz/Oe}$ is the electron gyromagnetic ratio and α is a phenomenological damping parameter. The value $\alpha = 0.1$ was chosen to give

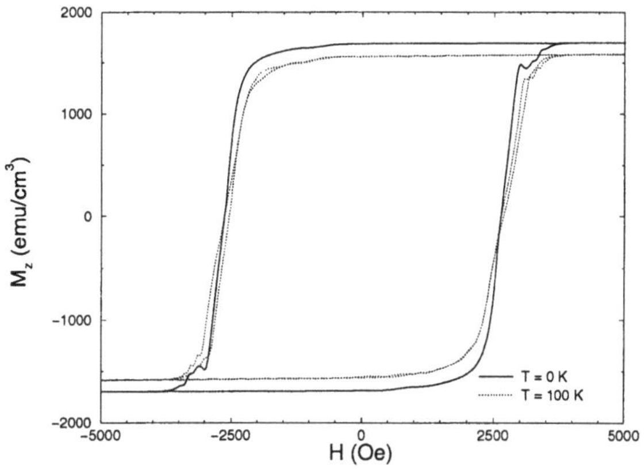

Fig. 3.1. Hysteresis loops of period 4 ns for flat-ended nanoscale pillars described in the text. One is at $T = 0\,\mathrm{K}$ (solid) and two at $T = 100\,\mathrm{K}$ (dotted)

an underdamped system. The saturation magnetization $M_s = 1700\,\mathrm{emu/cm^3}$ and the exchange length $\ell_x = 2.6\,\mathrm{nm}$ were chosen to match those of bulk iron. Numerical integration of (1) was carried out using a finite-differencing scheme with $\Delta r = 1.5\,\mathrm{nm}$ and $\Delta t = 5 \times 10^{-5}\,\mathrm{ns}$. The local field includes exchange interactions, dipole-dipole interactions, and thermal fluctuations. Details of the numerical approach are given in [6].

Here we consider nanomagnets with square cross sections of side 9 nm. One geometry has flat ends, and is 150 nm along the z-direction [7]. A simulated hysteresis loop at $T = 0\,\mathrm{K}$ with a period of 4 ns for this system is shown in Fig. 3.1. There is no crystalline anisotropy included in the model, but the shape anisotropy is quite strong, giving a coercive field of about 1980 Oe [6]. Even for these extremely fast loops, for a significant amount of time the magnet experiences a field just below the coercive field, and thermal fluctuations can carry the magnetization through the configuration associated with the energy barrier. Two hysteresis loops simulated at $T = 100\,\mathrm{K}$ are also shown in Fig. 3.1 to illustrate the extent of this effect.

A more dramatic way to see the effects of thermal fluctuations on the reversal of the magnetization is to hold the applied field fixed just below the coercive field. This traps the magnetization in a shallow metastable energy well, and thermal fluctuations become an essential part of the reversal process. Snapshots of the z-component of the magnetization separated by 0.05 ns are shown in Fig. 3.2a. The end caps associated with pole avoidance become strong in an antiparallel field, and thermal fluctuations affect their volume. The reversal starts when thermal fluctuations make one of these regions supercritical [7], after which it propagates at a constant rate towards the other end of the magnet. (A discussion of the saddle point associated with the nucleation process is found in [8].) Tapering the ends of the magnet reduces the pole-avoidance effects, and hence affects the end caps. However,

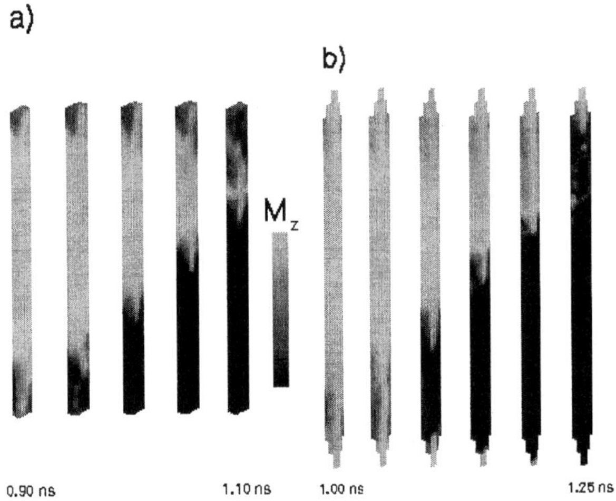

Fig. 3.2. Snapshots of the z-component of the magnetization, M_z, at equally spaced times for (**a**): pillars with square ends and $H_{\text{app}} = 1800\,\text{Oe}$, (**b**): pillars tapered at both ends and $H_{\text{app}} = 1975\,\text{Oe}$. All are identical cutaway views with only half of the pillar shown. Light shades indicate magnetization pointing in the unfavorable direction while the darkest shades indicate magnetization pointing in the favorable direction. Both simulations are for $T = 20\,\text{K}$

the switching is still characterized by nucleation at the ends; an example is shown in Fig. 3.2b for pillars 227 nm tall. The major effect we have observed of the tapering is an increase in the coercive field.

A simple model based on independent nucleation at the ends, followed by constant growth, allows calculation of the probability of not switching before time t to be [6,7]

$$P_{\text{not}}(t) = \begin{cases} 1 & t < t_0 \\ e^{-2\mathcal{I}(t-t_0)}\left[1 + 2\mathcal{I}(t-t_0)\right] & t_0 \le t < 2t_0 \\ e^{-2\mathcal{I}(t-t_0)}\left[1 + 2\mathcal{I}t_0\right] & 2t_0 \le t \end{cases}, \tag{3.2}$$

where t_0 is the earliest time at which a pillar can switch (when both ends start growing at $t = 0$) and $\mathcal{I}$ is the nucleation rate for each end. This form is compared to the simulation results for both pillars with flat and tapered ends in Fig. 3.3. The biggest obstacle to testing (3.2) is generating sufficient statistics at fields well below the coercive field [6,7].

In summary, thermal fluctuations and dipole-dipole interactions have been included in a three-dimensional model of single-domain nanoscale magnetic pillars. Simulated dynamics using the Landau–Lifshitz–Gilbert equation with thermal noise show that magnetization switching in the pillars occurs as the end caps nucleate, grow, and eventually coalesce. A simple model based on the thermally-activated nucleation of end caps for subcoercive fields gives a reasonable description of the simulation results for $P_{\text{not}}(t)$.

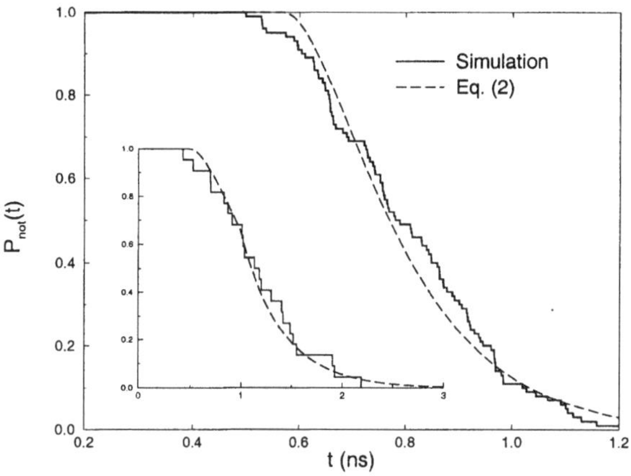

Fig. 3.3. Probability of not switching, $P_{\text{not}}(t)$, for 100 simulations of the flat-ended pillars at $H_{\text{app}} = 1800\,\text{Oe}$ and $T = 20\,\text{K}$, along with a fit to (3,2). The inset shows similar results for 22 trials with the tapered-end pillars at $H_{\text{app}} = 1975\,\text{Oe}$ and $T = 20\,\text{K}$

Acknowledgement

Supported by NSF Grant Nos. DMR-9871455 and DMR-0120310, U.S. DOE, NERSC, and by FSU/SCRI, FSU/CSIT, and FSU/MARTECH.

References

1. P.A. Rikvold, M.A. Novotny, M. Kolesik, H.L. Richards: "Nucleation Theory of Magnetization Reversal in Nanoscale Ferromagnets". In: *Dynamical Properties of Unconventional Magnetic Systems*, ed. by T. Skjeltorp, D. Sherrington (Kluwer, Dordrecht 1998) p. 307
2. P.A. Rikvold, G. Brown, S.J. Mitchell, M.A. Novotny: "Dynamics of Magnetization Reversal in Models of Magnetic Nanoparticles and Ultrathin Films". In: *Nanostructured Magnetic Materials and their Applications*, ed. by D. Shi, B. Aktas, L. Pust, F. Mikailov (Springer, Berlin Heidelberg New York 2002) p. 184
3. S. Wirth, M. Field, D.D. Awschalom, S. von Molnár: Phys. Rev. B **57**, R14028 (1998); J. Appl. Phys. **85**, 5249 (1999)
4. W.F. Brown: Phys. Rev. **130**, 1677 (1963)
5. A. Aharoni: *Introduction to the Theory of Ferromagnetism* (Clarendon, Oxford 1996)
6. G. Brown, M.A. Novotny, P.A. Rikvold: Phys. Rev. B **64**, 134422 (2001)
7. G. Brown, M.A. Novotny, P.A. Rikvold: J. Appl. Phys. **87**, 4792 (2000)
8. G. Brown, M.A. Novotny, P.A. Rikvold: "Determining the Saddle Point in Micromagnetic Models of Magnetization Switching". In: *Computer Simulation Studies in Condensed Matter Physics XV*, ed. by D.P. Landau, S.P. Lewis, H.-B. Schüttler (Springer, Berlin Heidelberg New York 2003) (This volume)

4 Determining the Saddle Point in Micromagnetic Models of Magnetization Switching

G. Brown[1,2], M.A. Novotny[3], and P.A. Rikvold[2,4]

[1] Center for Computational Sciences, Oak Ridge National Laboratory, P.O.Box 2008 Mail Stop 6114, Oak Ridge, TN 37831-6114, USA
[2] School of Computational Science and Information Technology, Florida State University, Tallahassee, FL 32306-4120, USA
[3] P.O. Box 5167, Department of Physics and Astronomy, Mississippi State University,
Mississippi State, MS 39762-5167, USA
[4] Center for Materials Research and Technology and Department of Physics, Florida State University, Tallahassee, FL 32306-4351, USA

Abstract. A numerical model of single-domain nanoscale iron magnets fabricated using scanning-microscope-assisted chemical vapor deposition is simulated using finite-temperature micromagnetics. A projective-dynamics method is used to determine the magnetization at the saddle point as a function of temperature. This magnetization is found to decrease linearly as the temperature is raised.

An interesting problem in nonequilibrium statistical mechanics, with numerous applications in condensed-matter physics and materials science, is how a system approaches the global minimum of the free energy. One simple, but common, situation is that of a local free-energy minimum which is separated from the global minimum by a free-energy barrier. In other words, any path between the metastable local minimum and the equilibrium global minimum involves an initial increase in the free energy. The most probable path between the minima involves the minimum free energy increase since spontaneous increases in free energy are improbable. The maximum free energy along that most probable path corresponds to the saddle point. Often, the behavior of the nonequilibrium dynamics are dominated by properties near the minima and the saddle point. While the minima can be determined by examining histograms of the state of the system, the saddle point is much harder to determine. In this paper we present a technologically important example of finding the saddle point using a projective-dynamics technique.

The results described here are for a numerical model of single-domain nanoscale iron magnets that have been fabricated using chemical vapor deposition directed by a scanning-tunneling microscope [1]. These magnetic pillars have cross-sectional dimensions on the order of 10 nm and extend on the order of 100 nm perpendicular to the substrate. The magnetic pillars are modeled by a one-dimensional array of magnetization density vectors $\boldsymbol{M}(\boldsymbol{r})$

with fixed length M_s. Each vector precesses around a local field $\boldsymbol{H}(\boldsymbol{r})$ according to the Landau–Lifshitz–Gilbert equation [2,3]

$$\frac{\mathrm{d}\boldsymbol{M}(\boldsymbol{r})}{\mathrm{d}t} = \frac{\gamma_0}{1+\alpha^2}\boldsymbol{M}(\boldsymbol{r}) \times \left[\boldsymbol{H}(\boldsymbol{r}) - \frac{\alpha}{M_s}\boldsymbol{M}(\boldsymbol{r}) \times \boldsymbol{H}(\boldsymbol{r})\right], \tag{4.1}$$

where the electron gyromagnetic ratio $\gamma_0 = 1.76 \times 10^7\,\mathrm{Hz/Oe}$. The phenomenological damping parameter $\alpha = 0.1$ was chosen to give underdamped dynamics. Properties of bulk iron were assumed with the saturation magnetization $M_s = 1700\,\mathrm{emu/cm^3}$ and the exchange length $\ell_x = 2.6\,\mathrm{nm}$. The model pillar has a square cross section with area $4\ell_x^2$, and is $34\ell_x$ long. Details of the numerical approach, including the inclusion of thermal fluctuations in the local field, are given in [4].

These nanopillars have a strong shape anisotropy, and the magnetization is most favorably oriented along the long axis of the magnet, which is taken to be the z direction. The nanomagnets are prepared in a metastable state via equilibration in an externally-applied magnetic field $+H_0\hat{z}$, which is subsequently varied rapidly in magnitude (without changing its orientation) to $-H_0\hat{z}$. If the field H_0 is less than the coercive field, here $H_c \sim 1500\,\mathrm{Oe}$ [4], a free-energy barrier separates the positively-oriented magnetization from the equilibrium negative orientation. Previous simulations [4] suggest that the saddle point in these pillars corresponds to the nucleation of a region of reversed magnetization at one of the ends of the pillar. The nucleated region grows until the entire magnetization is reversed.

A useful technique for determining the magnetization corresponding to the free-energy maximum of the barrier, also called the saddle point, is the projective-dynamics technique [5–7]. The essence of this technique is projecting the original dynamics described in a high-dimensional phase space onto a probabilistic dynamic in a phase space of much lower dimension. A specific example is helpful. The dynamics of an Ising model can be projected onto the number of overturned spins, a measure of the global magnetization. The dynamics of the one-dimensional model are then described in terms of the probability that the number of overturned spins increases or decreases. Thinking in terms of the number of spins in the stable orientation, these probabilities are referred to as the probability of growing (P_{grow}) and shrinking (P_{shrink}), respectively [5].

Projective dynamics for the model nanopillars is more complicated than that for the Ising model for two reasons. First, the magnetization is a continuous, three-dimensional variable. This is handled by projecting the z-component of the global magnetization into uniformly sized bins [7]. Then P_{grow}, the probability of the region of stable magnetization "growing," is the probability that during an integration step the magnetization moves into a bin corresponding to a smaller magnetization. Similarly, P_{shrink} is the probability that the magnetization moves into a bin corresponding to a larger magnetization. The second complication involves the persistent, subcritical regions of reduced magnetization that develop at each end of the pillar. To

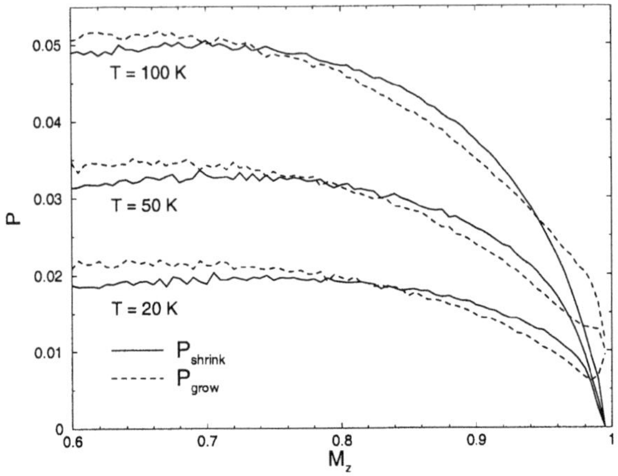

Fig. 4.1. Probability of "growing" and "shrinking" as functions of the magnetization of one half of the pillar for three temperatures: 100 K, 50 K, and 20 K. The left-most crossing shown for a given temperature indicates the saddle point, while the right-most crossing indicates the metastable local free-energy minimum. The crossings associated with the equilibrium global free-energy minimum lie near −1 and are not shown

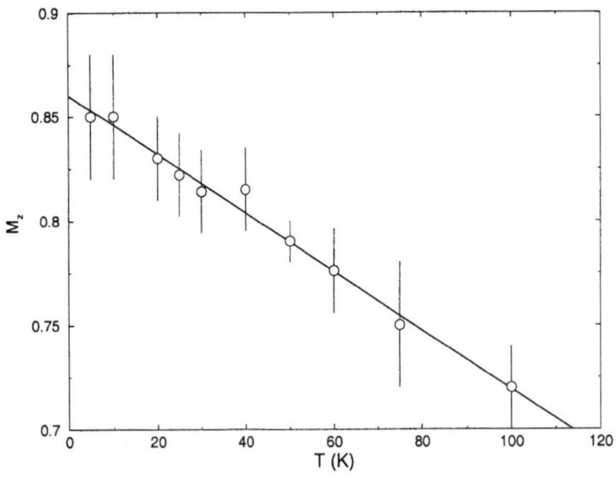

Fig. 4.2. Magnetization along the pillar, M_z at the saddle point as a function of temperature (circle). The magnetization at the saddle point decreases with increasing temperature, and the solid line is a least-squares fit with intercept at 0.860. The error bars are estimated from the range of M_z where P_{grow} and P_{shrink} are nearly equal

accomodate this, during the projective-dynamics analysis the pillar is divided into its top and bottom parts with the normalized, "global" magnetization M_z of each half considered separately.

The measured P_{grow} and P_{shrink} at $H_0 = 1000\,\text{Oe}$ are shown in Fig. 4.1 for temperatures of 20 K, 50 K, and 100 K. For each temperature the results represent averaging over more than 10^9 integration steps and on the order of 10^4 switches. Consider the probabilistic dynamics at $M_z = 0.85$ for $T = 100\,\text{K}$. Here $P_{\text{shrink}} > P_{\text{grow}}$ and on average the magnetization moves to the right. For $M_z = 0.98$ the situation is reversed, and on average the magnetization moves to the left. The crossing point where $P_{\text{shrink}} = P_{\text{grow}}$ corresponds to a locally stable fixed point since, on average, here the magnetization moves towards the crossing. In this case the crossing is the metastable local free-energy minimum. For the crossing near $M_z = 0.74$ the magnetization moves away from the point on average, and it corresponds to an unstable local maximum in the free energy. This is the saddle point. Measuring the value of M_z for which $P_{\text{shrink}} = P_{\text{grow}}$ is a convenient method for finding the saddle point.

From Fig. 4.1 it is obvious that M_z at the saddle point is different for magnetization reversal at different temperatures. The value of M_z at the saddle point is presented vs temperature in Fig. 4.2. The point where $P_{\text{shrink}} = P_{\text{grow}}$ is estimated using lines fit via a least-squares method to each probability in the region near the crossing. The error bars are taken large enough to include nearly all M_z for which the difference in the probabilities is less than the fluctuations in their estimates.

There is a clear linear trend in M_z at the saddle point with respect to the temperature. The solid line in Fig. 4.2 is an unweighted least-squares fit to the data. This fit indicates that $M_z(T \to 0\,\text{K}) = 0.860$ at the saddle point. The dependence on temperature is quite strong, since $M_z(100\,\text{K}) = 0.72$.

In summary, micromagnetic simulations at finite temperature have been used to investigate magnetization reversal in single-domain nanoscale magnets. The saddle point in the free energy was determined using the projective-dynamics technique, and the magnetization at the saddle point was found to depend linearly on the temperature for all the temperatures investigated, $T \leq 100\,\text{K}$.

Acknowledgements

This work was supported in part by the Laboratory Directed Research and Development Program of Oak Ridge National Laboratory, managed by UT-Battelle, LLC for the U.S. Department of Energy under Contract No. DE-AC05-00OR22725, and by the Ames Laboratory, which is operated for the U.S. Department of Energy by Iowa State University under Contract No. W-7405-82. This work was also supported by NSF grant No. DMR-0120310 and by FSU/CSIT. Computer resources were provided by the FSU Deptartment of Physics.

References

1. S. Wirth, M. Field, D.D. Awschalom, S. von Molnár: Phys. Rev. B **57**, R14028 (1998); J. Appl. Phys. **85**, 5249 (1999)
2. W.F. Brown: Phys. Rev. **130**, 1677 (1963)
3. A. Aharoni: *Introduction to the Theory of Ferromagnetism* (Clarendon, Oxford 1996)
4. G. Brown, M.A. Novotny, P.A. Rikvold: Phys. Rev. B **64**, 134422 (2001)
5. M. Kolesik, M.A. Novotny, P.A. Rikvold: Phys. Rev. Lett. **80**, 3384 (1998)
6. M.A. Novotny: Int. J. Mod. Phys. C **10**, 1483 (2000); M.A. Novotny: *Annual Reviews of Computational Physics IX*, D. Stauffer (Ed.) (World Scientific, Singapore 2001) p. 153
7. S.J. Mitchell, M.A. Novotny, J.D. Muñoz: Int. J. Mod. Phys. C **10**, 1503 (2000)

5 Macroscopic Magnetization Jumps in Strongly Frustrated Quantum Spin Lattices

J. Schulenburg and J. Richter

Otto-von-Guericke-Universität Magdeburg, 39016 Magdeburg, Germany

Abstract. We discuss typical features of magnetization curves of quantum antiferromagnets. For a class of frustrated spin lattices including the kagomé lattice we find a linear dependence of the energy on the quantum number of total S^z. This linearity leads to a jump in the magnetization versus field curve near saturation and is connected with a set of exact eigenstates consisting of independent localized magnons. The observed jump is a true macroscopic quantum effect which is also of experimental relevance since it occurs in many different lattices.

5.1 Introduction

Frustrated quantum spin systems exhibit rich physics including the possibility of collinear and noncollinear ground state (GS) ordering, gapped quantum disordered GS as well as quantum critical behavior. Many experimentally studied systems are well described by the antiferromagnetic quantum Heisenberg model. In spite of its simple form exact solutions are very rare. Recently much attention has been paid to the GS properties of the Heisenberg antiferromagnet (AFM) in magnetic field. In recent measurements of magnetization curves of the two-dimensional gapped quantum magnet $SrCu_2(BO_3)_2$ the existence of three plateaus at $m = 1/8$, $1/4$ and $1/3$ of full magnetization has been reported [1]. This observation is in good agreement with theoretical predictions based on the corresponding quantum Heisenberg model [2]. Besides plateaus also jumps in the magnetization curve can occur. This discontinuity is called "spin-flop" or "metamagnetic" transition. Metamagnetism is observed in classical as well as in quantum systems and can be attributed to anisotropic terms in the Hamiltonian.

After a short illustration of the basic features of magnetization curves in Sect. 5.2 we discuss in this paper a new type of a macroscopic magnetization jump. This new kind of jump is a pure quantum effect and appears in strongly frustrated AFM even without any anisotropic terms in the Hamiltonian.

We start with the XXZ-Heisenberg AFM in a magnetic field h

$$\hat{H} = \sum_{\langle ij \rangle} J_{ij} \left\{ \Delta \hat{S}_i^z \hat{S}_j^z + \frac{1}{2}\left(\hat{S}_i^+ \hat{S}_j^- + \hat{S}_i^- \hat{S}_j^+ \right) \right\} - h\hat{S}^z \,; \quad J_{ij} \geq 0 \,, \tag{5.1}$$

where the sum runs on nearest neighbors. We focus on the isotropic case $\Delta = 1$ but our considerations are not restricted to that case. We emphasize that the

z-component of the total spin $\hat{S}^z = \sum_i \hat{S}_i^z$ commutes with the Hamiltonian. The quantum number M of $\hat{S}^z$ determines the normalised magnetization $m = M/(sN) = M/M_{\rm max}$, where N is the number of spins and s the spin quantum number of a single spin. Maximal quantum fluctuations belong to $s = 1/2$ whereas $s \to \infty$ corresponds to the classical limit of the model.

5.2 General Features of Magnetization Curves

For a better understanding of the discussions given in the following sections we present in this section a brief introduction into some interesting features of magnetization curves. For more details see e.g. [3].

The zero-field energy per site $e = E/N$ of (5.1) depends on the quantum number $M = mM_{\rm max}$. Hence we get the magnetization versus field curve $m(h)$ by minimizing $e(m) - mh$, which leads to $h(m) = \partial e/\partial m$. In a uniform unfrustrated AFM with a collinear long-range ordered GS $e(m)$ is almost quadratic $e(m) \sim m^2/(2\chi)$ and the magnetization is almost linear. Typical examples for that are the Heisenberg AFM on the square or the honeycomb lattice. In frustrated spin systems the magnetization process can be more complicated and the $e(m)$ curve may differ from the almost quadratic dependence. Discontinuities in the slope of $e(m)$ lead to a plateau and concave parts in the $e(m)$ curve lead to metamagnetic transitions (jumps) in $m(h)$.

The Heisenberg AFM on the triangular and the kagomé lattice are examples for a plateau at $m = 1/3$. The origin of this plateau can be understood in terms of the Ising part of the Hamiltonian (5.1). For the Ising AFM on the triangular and the kagomé lattice e depends linearly on m but the slope of $e(m)$ changes at $m = 1/3$. Including the transversal spin interactions in (5.1) the linear curves become almost quadratic but the change in the slope at $m = 1/3$ survives [4]. Therefore these plateaus can be considered as being of (semi-)classical origin.

The occurence of a metamagnetic transition in quantum spin systems was for instance found in the frustrated J_1–J_2 XXZ chain [5] for highly anisotropic exchange ($\Delta \approx -0.5$). Here we have a competition of ferro- and antiferromagnetic GS and $e(m)$ becomes concave near the saturation of magnetization.

5.3 Magnon Traps in the Kagomé and Other Strongly Frustrated Lattices

In more detail we will discuss the spin-half Heisenberg AFM on the kagomé lattice. This system has attracted much attention because there are strong arguments in favor of a quantum disordered GS [6–8], whereas the Heisenberg AFM on the triangular lattice is Néel ordered.

The energy versus magnetization curve $e(m)$ for the $s = 1/2$ Heisenberg AFM on a finite kagomé lattice with $N = 36$ sites is shown in Fig. 5.1 (left

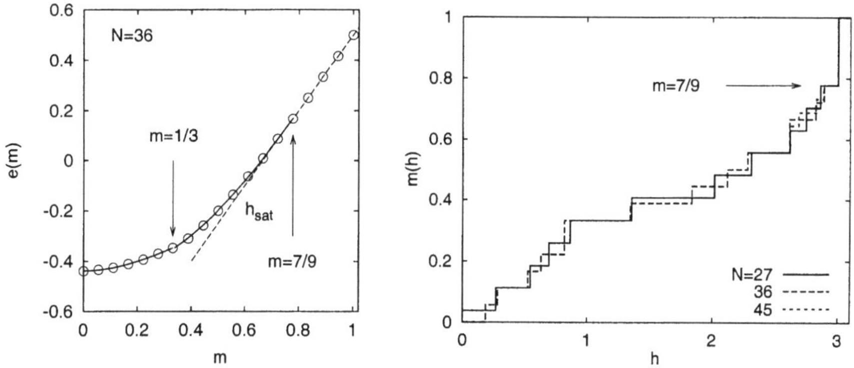

Fig. 5.1. The $e(m)$ curve for the isotropic spin-half Heisenberg AFM on the kagomé lattice has a discontinuous slope at $m = 1/3$ and an exactly linear part above $m = 7/9$ (dashed line). The points in the left figure show the lowest states for the $N = 36$ system. The curve is a guide to the eyes. On the right figure the jump in the magnetization curve is shown for finite systems of various size

side). As discussed qualitatively in the last section there is a discontinuity in the slope of $e(m)$ at $m = 1/3$. But besides this characteristic feature (which is also present for the triangular lattice) Fig. 5.1 shows another important feature in $e(m)$ (which is not present for the triangular lattice), namely an exactly linear part above $m = 7/9$. This linear dependence of e on m causes a jump from $m = 7/9$ to full magnetization $m = 1$ (Fig. 5.1, right side).

To explain the origin of the linear part of $e(m)$ we start from the fully polarized ferromagnetic state with $m = 1$ realized in strong magnetic fields $h \geq h_{sat}$. We consider the fully polarized state as the vacuum state $|0\rangle$. The low-lying excitations for $h \geq h_{sat}$ are one-magnon states $|1\rangle = \frac{1}{c}\sum_i^N a_i S_i^- |0\rangle$. We have three one-magnon bands $w(k)$ according to the three sites in the unit cell of the kagomé lattice. The lowest one-magnon band is completely flat and its energy is $w = -2sJ(1 + 2\Delta)$. This degeneracy of the extended one-magnon states allows to construct localized magnon states.

A localized magnon state on the kagomé lattice is illustrated in Fig. 5.2. The corresponding coefficients a_i of the one-magnon state $|1\rangle = \frac{1}{c}\sum_i^N a_i S_i^- |0\rangle$ are zero for all lattice sites except the six sites $i \in L$ around a hexagon. It can be easily shown that the one-magnon state with alternating coefficients $a_i = (-1)^i$ is an eigenstate of the Heisenberg AFM on a hexagon. This eigenstate for a hexagon is also an eigenstate of the whole kagomé lattice. This fact is connected with the triangles surrounding the hexagon leading to a cancellation of terms $S_{l_1}^+ S_j^-$ and $S_{l_2}^+ S_j^-$ if $a_{l_1} = -a_{l_2}$, where l_1 and l_2 label neighboring sites on the hexagon and j labels the site of the corresponding triangle outside the hexagon. As a consequence the magnon is trapped (localized) on the hexagon.

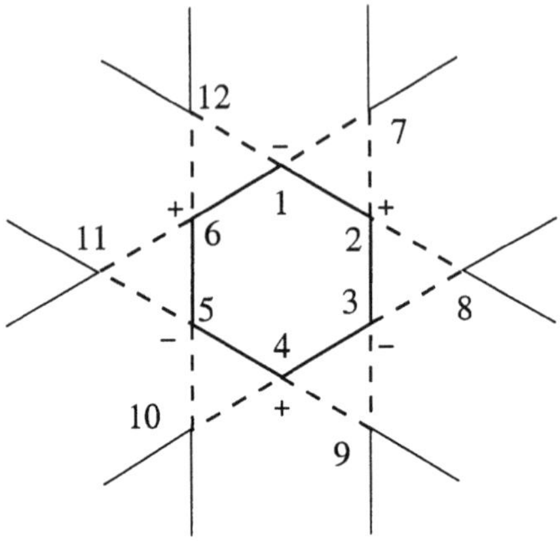

Fig. 5.2. A magnon trapped on a hexagon surrounded by triangles (dashed)

Clearly we can put further trapped magnons on hexagons separated from each other by at least a minimum Manhattan distance of two, this way preventing interference of the localized magnons. By putting a further trapped magnon on the lattice we lower the magnetization by one and the zero-field energy by the one-magnon energy w leading to a linear dependence of $e(m)$ and finally to the observed magnetization jump. The highest density of independent magnons is reached if every third hexagon carries a local magnon. In this case $N/9$ local magnons form a crystal like structure. The magnetization is therefore reduced by $\delta m = (N/9) \times (1/sN)$ in this state, which defines the height of the jump.

The general condition for trapping a magnon on a local region L of the lattice is [9]

$$\sum_{l \in L} a_l J_{lk} = 0 \quad \forall \quad k \in R\,, \tag{5.2}$$

where R are the lattice sites outside the local region L. This condition leads to quite a lot of lattice variants to trap a magnon. In its simplest form such lattices have polygons with even number of sites and are surrounded by triangles on every or every second edge. Further examples are the checkerboard lattice, the squagomé lattice [10], the three-dimensional pyrochlore lattice, the one-dimensional sawtooth chain [11], the kagomé chain [9] or finite systems describing giant magnetic molecules [10]. We notice, that localized magnon states in a ferromagnetic background define a class of exact product eigenstates of (5.1) on these lattices.

5.4 Summary

We have shown, that special geometries of strongly frustrated spin lattices, for instance even polygons surrounded by triangles, can lead to a macroscopic jump in the magnetization curve to full magnetization $m = 1$. The height of the jump is inverse proportional to the spin quantum number s. Therefore the jump vanishes in the classical limit. Since the decoupling condition (5.2) is independent of Δ, the existence of the jump is not connected with a special anisotropy like the metamagnetic transition reported e.g. in [5]. From the decoupling condition (5.2) it is possible to create new lattices, which show this magnetization jump too. All of these lattices are necessarily strongly frustrated systems. The reported jump represents a genuine macroscopic quantum effect which is also of possible experimental relevance since it occurs in many well-known models like the kagomé lattice.

References

1. K. Onizuka, H. Kageyama, Y. Narumi, K. Kindo, Y. Ueda, T. Goto: J. Phys. Soc. Jap. **69**, 1016 (2000)
2. G. Misguich, T. Jolicoeur, S.M. Girvin: Phys. Rev. Lett. **87**, 97203 (2001)
3. C. Lhuillier, G. Misguich, cond-mat/0109146
4. A. Honecker: J. Phys. Condens. Matter **11**, 4697 (1999)
5. Gerhardt, K.-H. Mütter, H. Kröger: Phys. Rev. B **57**, 11504 (1998)
6. P. Lecheminant, B. Bernu, C. Lhuillier, L. Pierre, P. Sindzingre: Phys. Rev. B **56**, 2521 (1997)
7. C. Waldtmann, H.-U. Everts, B. Bernu, C. Lhuillier, P. Sindzingre, P. Lecheminant, L. Pierre: Eur. Phys. J. B **2**, 501 (1998)
8. F. Mila: Phys. Rev. Lett. **81**, 2356 (1998)
9. J. Schulenburg, A. Honecker, J. Schnack, J. Richter, H.-J. Schmidt et al.: Phys. Rev. Lett. **88**, 167207 (2002)
10. J. Schnack, H.-J. Schmidt, J. Richter, J. Schulenburg: Eur. Phys. J. B **24**, 475 (2001)
11. T. Nakamura, K. Kubo: Phys. Rev. B **53**, 6393 (1996)

6 Nonequilibrium Relaxation Study on the Ferromagnetic Transition of Ising Spin Glass Model

N. Ito[1] and Y. Ozeki[2]

[1]Department of Applied Physics, School of Engineering, The University of Tokyo, Hongo, Bunkyo-ku, Tokyo 113-8656, JAPAN
[2]Department of Physics, Tokyo Institute of Technology, Oh-okayama, Meguro-ku, Tokyo 152-8551, JAPAN

Abstract. Ferromagnetic transition of the $\pm J$ Ising model is studied simulationally for the cubic lattice. Nonequilibrium relaxation method is used to locate the transition point and to estimate the values of exponents. The dynamical exponent z increases smoothly when the density of ferromagnetic interaction, p, is decreased from 1. The static exponents are changed discontinuously when p is decreased from 1, and they stay in a universality class when $p_{\rm mc} < p < 1$, where $p_{\rm mc}$ denotes the value of multicritical density. At $p = p_{\rm mc}$, the static exponents are different from those in $p_{\rm mc} < p < 1$ and $p = 1$.

6.1 Introduction

Much effort has been devoted to elucidate the nature and the behavior of disordered systems. Disorder is typically introduced by impurities and defects and it is theoretically treated using the model with quenched random variables like site dilution and random coupling constant. One basic question about the disordered system is whether the thermodynamic state is different from that of pure system or not [1].

Disorder gives rise to intrinsic state, for example, the spin-glass phase and the localized state. Even the state which appears without disorder, for example, the ferromagnetic(FM) phase, is sometimes influenced by disorder and the universality class of transition is altered. The condition that such alteration happens is that the specific heat exponent, α, of pure system is positivc [2]. This criterion is originally derived by considring that the fluctuation of transition temperature due to the fluctuation of quenched disorder in each sample is negligible or not. Assuming that the transition temperature depends linearly on the density of disordered interaction, it will be the order of $\xi^{-d/2} \sim \epsilon^{d\nu/2}$, where d, ξ, ν and ϵ denote the lattice dimensionality, correlation length, its exponent and $|T - T_c|$. Therefore it will be irrelevant when $\epsilon^{d\nu/2}$ is negligible compared with ϵ, that is, $d\nu/2 > 1$. This inequality implies $\alpha < 0$ using the hyperscaling relation $\alpha = 2 - d\nu$. Inversely it will be relevant when $\alpha > 0$. This criterion is often refered as the Harris criterion [2]. The derivation of this criterion is phenomenological, and it is confirmed rigorously that ν of disordered system is larger than $2/d$ [3]. So far we have assumed that the density of randomness fluctuates because each random bond or site is determined independently. Such way of random averaging is sometimes called as grand canonical averaging. It is shown using renormalization group method that the Harris criterion will be valid in case of canonical averaging where the total density of randomness is fixed [4].

On three-dimensional lattice, α of the purely FM Ising model is estimated around 0.12. Therefore the Harris criterion suggests that the universality class of random Ising FM transition is different from one of the pure system on three-dimensional lattice. The next question is what kind of universality class appears in the disordered system when α of the pure system is positive. The renormalization group analysis suggested that the disordered system belongs to the same universality class driven by so-called the random fixed point [5,6]. Estimated values of exponents of three-dimensional random Ising FM transition is $\nu \approx 0.67$ and $\beta \approx 0.35$ and so on [7], and they are consistent with the experimental results [8].

Computer simulation has also been playing an active role in the study on disordered systems. It is applied to test the Harris criterion and the random fixed point. For this purpose, bond- or site- diluted Ising model on cubic lattice is the most suitable system [9,10], and the recent studies concluded that the simulation results are consistent with them [11–13]. Spin-glass model is also a typical random system and it is interesting to study whether its FM transition is described by the above scenario. Recent Monte Carlo renormalization group study suggested that it is the case [14].

Standard strategy, to which we refer here as equilibrium method, of statistical mechanical simulation is to evaluate the equilibrium values of physical quantities and to analyze them. Another strategy named the nonequilibrium relaxation(abbreviated as NER) method has been proposed and clutivated, which analyze nonequilibrium relaxation process from an initial state to the equilibrium state [15–20]. The initial equilibration is just discarded when the equilibrium method is used, but the relaxation time of disordered system and/or in the critical region often becomes very long. So the NER method usually becomes advantageous. NER method has turned out to be useful to study various kinds of phases and transitions, for example, standard transitions, Kosterlitz–Thouless transition and quantum systems. It has been shown that this NER method is especially useful to study the disordered systems [16,21–23,25,26].

In this article, the result of nonequilibrium relaxation study on the FM transition of $\pm J$ Ising model on cubic lattice is given. The model is defined by the interaction energy,

$$E = -\sum_{<i,j>} J_{ij}\sigma_i\sigma_j, \quad \sigma_i = \pm 1\,, \quad J_{ij} = \begin{cases} J & \text{probability } p \\ -J & \text{probability } 1-p \end{cases}, \tag{6.1}$$

where p denotes the density of FM interaction, and J is positive. Dimensionless inverse temperature $K = \beta J = J/T$ is used here to denote the temperature T. Cubic lattice are mainly treated.

6.2 Phase Diagram of $\pm J$ Ising Model

In the $p - T$ plane, the $\pm J$ Ising model has paramagentic(PM), FM and spin-glass(SG) phases on cubic lattice. The Nishimori line defined by $\mathrm{e}^{2K} = p/(1-p)$ plays a special role [27]. This model is solved exactly along this line using the gauge-transformation technique. Along this line, the specific heat does not show divergence. The multicritical point is considered to be located on this line. If there is no spontaneous magnetization at p and $T = T_p$ on this line, it is shown that there is no spontanous magnetization at any $T < T_p$ [27].

The transition point is located by observing the behavior of the order-parameter NER function $f_m(t)$ starting from the completely ordered initial state. The local exponent $\lambda_m(t)$ of $f_m(t)$ is defined by

$$\lambda_m(t) = -\frac{\mathrm{d}\log f_m(t)}{\mathrm{d}\log t} . \tag{6.2}$$

The estimated values of $\lambda_m(t)$ is plotted versus $1/t$. On this plot, the PM and FM phases are characterized by turn-up and down behavior in the limit of $1/t \to 0$ [16]. At the second-order transition point, this $\lambda_m(t)$ is extrapolated to a positive value λ_m which is equal to $\beta/z\nu$ where β and z denote the magnetization and dynamical exponents, respectively [15].

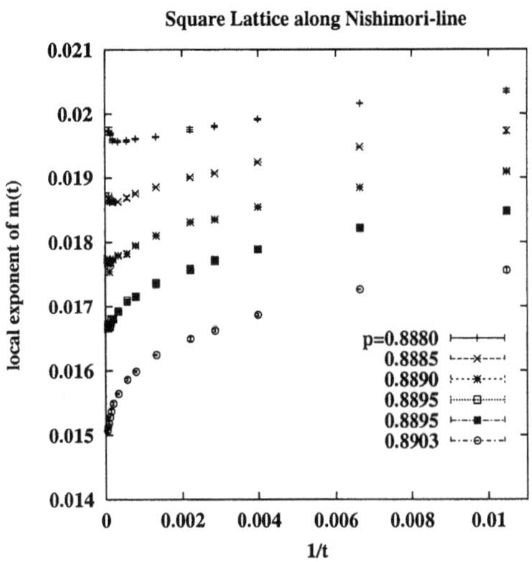

Fig. 6.1. The estimated values of $\lambda_m(t)$ of square $\pm J$ Ising model along the Nishimori-line is plotted versus $1/t$ at some values of p

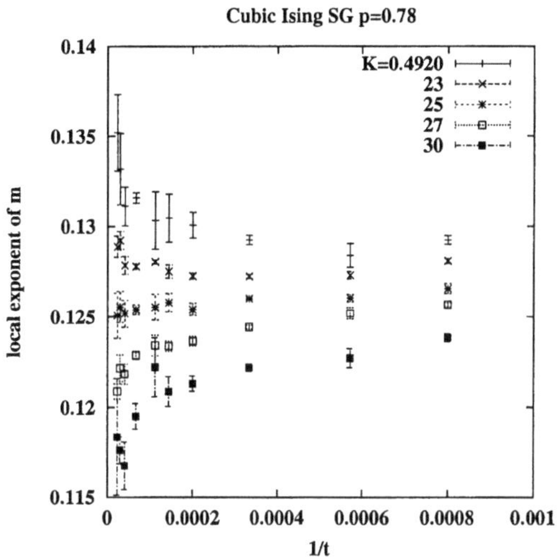

Fig. 6.2. The estimated values of $\lambda_m(t)$ of cubic $\pm J$ Ising model at $p = 0.78$ is plotted. $K = 0.4923$ and 0.4927 show PM and FM behaviors, respectively

Table 6.1. The estimated values of transition (inverse) temperature K_c of $\pm J$ Ising model on cubic lattice with NER method are listed. $T_{\text{max.}}$ and $N_{\text{typ.}}$ denote the maximum time in MCS from all-up initial state and typical number of samples

p	K_c	typical lattice	$T_{\text{max.}}$	$N_{\text{typ.}}$
1.00 [19]	0.2216545(15)	$501 \times 501 \times 502$	2,000	200,000
0.90	0.285285(5)	$451 \times 451 \times 452$	20,000	6,400
0.80	0.4251(1)	$451 \times 451 \times 452$	10,000	3,840
0.78	0.4925(2)	$349 \times 349 \times 350$	50,000	11,000
$p_{\text{mc}} = 0.7673(3)$ [21]	0.5966(8)	$161 \times 161 \times 162$	20,000	1,000

The results of square $\pm J$ Ising model along Nishimori-line is shown in Fig. 6.1. Simulations up to 50 000 Monte Carlo Sweeps (MCS) with two-sublattice local dynamics were made. Lattices up to $10\,001 \times 10\,000$ were used. The curve of $p = 0.8885$ shows paramagnetic turn-up, and that of $p = 0.8903$ shows FM turn-down. So the transition point is located to be $p_{\text{mc}} = 0.8894(9)$.[1] The exponent $\lambda = \beta/z\nu$ is estimated to be 0.017(1). The transition temperatures of the cubic $\pm J$ Ising model are also estimated at several concentration p. The results are listed in Table 6.1. Figure 6.2 shows the local exponents at $p = 0.78$.

[1] Previous estimation of p_{mc} from the simulation up to 750 MCS [21] is refined here observing the longer time behavior.

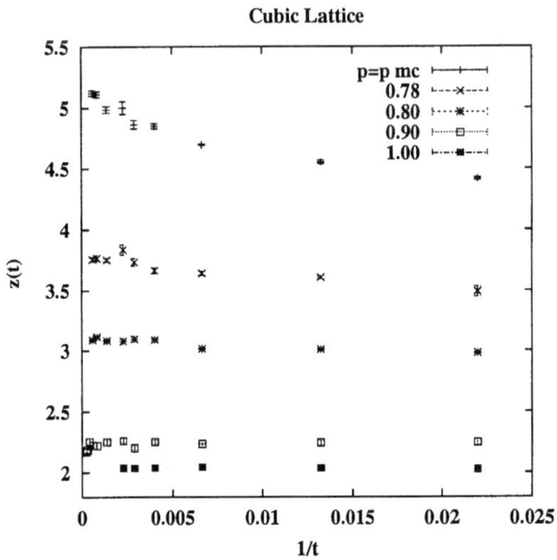

Fig.6.3a

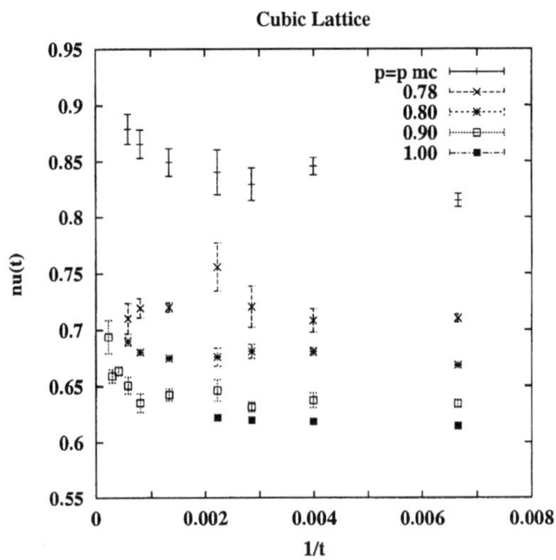

Fig.6.3b

6.3 Exponent

Exponents are estimated from the NER of fluctuations of physical quantities at the transition point [17,19,23,24], which are defined by

$$f_{mm}(t) = L^d \left[\frac{\langle m(t)^2 \rangle}{\langle m(t) \rangle^2} - 1 \right] \quad \text{and} \quad f_{me}(t) = L^d \left[\frac{\langle m(t)e(t) \rangle}{\langle m(t) \rangle \langle e(t) \rangle} - 1 \right] ,(6.3)$$

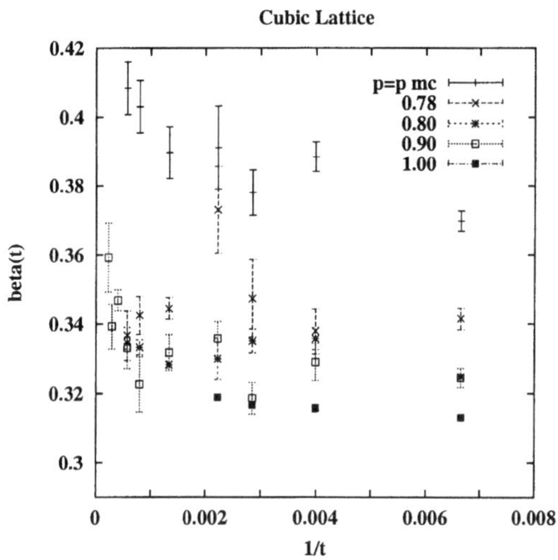

Fig.6.3c

Fig. 6.3. Estimated values of the local exponents (**a**) $z(t)$, (**b**) $\nu(t)$ and (**c**) $\beta(t)$ for cubic $\pm J$ Ising model at $p = 1.0, 0.9, 0.8, 0.78, p_{\rm mc}$ are shown. Typical conditions of Monte Carlo simulation were 1.8×10^6 samples up to 2000 MCS for $151 \times 151 \times 152$ lattice

where $m(t)$ and $e(t)$ denote the values of magnetization and energy per spin, respectively, after t MCS from all-up initial configuration. Using their local exponents defined by

$$\lambda_I(t) = \frac{\mathrm{d}\log f_I(t)}{\mathrm{d}\log t} \quad (I = mm \text{ and } me)\,, \tag{6.4}$$

the local exponents converging to the conventional static and dynamic exponents in the limit of $1/t \to 0$ are defined by

$$z(t) = \frac{d}{\lambda_{mm}(t)}\,, \nu(t) = \frac{\lambda_{mm}(t)}{d \cdot \lambda_{me}(t)} \text{ and } \beta(t) = \frac{\lambda_m(t)}{\lambda_{me}(t)}\,. \tag{6.5}$$

At the transition points listed in Table 1, estimated values of these local exponents are given in Fig. 6.3. From Fig. 6.3a, it is observed that the dynamical exponent z is 2.2(1) at $p = 0.90$, 3.1(1) at $p = 0.80$, 3.8(1) at $p = 0.78$, and 5.2(1) at $p = p_{\rm mc}$. At $p = 1.00$, it was estimated to be 2.055(10)[17], From Fig. 6.3b, it is observed that value of ν at $p = p_{\rm mc}$ is different from ones at p in $1 > p > p_{\rm mc}$. At $p = p_{\rm mc}$, ν is estimated to be 0.90(2), but the local exponents for $p = 0.78 \sim 0.90$ are all extrapolated to the same value, 0.67(2). From Fig. 6.3c, the similar behavior is observed in β as in ν. The estimated value of β at $p = p_{\rm mc}$ is 0.42(2), and it is 0.34(1) at p in $1 > p > p_{\rm mc}$. These values, $\nu = 0.67(2)$ and $\beta = 0.34(1)$, in $1 > p > p_{\rm mc}$ are different from those of purely FM system ($p = 1$), that is, $\nu = 0.635(5)$ and $\beta = 0.325(5)$ [17]. Estimated values of static exponents are listed in Table 6.2.

Table 6.2. NER estimations for static exponents are summarized

p	$p = 1$	$1 > p > p_{\rm mc}$	$p_{\rm mc}$
ν	0.635(5)	0.67(2)	0.90(2)
β	0.325(5)	0.34(1)	0.42(2)

6.4 Summary

Ferromagnetic transition of the $\pm J$ Ising model on cubic lattice is studied by using the NER method. The transition points are located by analyzing the local exponent of the magnetization NER function from all-up initial configuration. At each transition point, the dynamic and static exponents are estimated by using the NER functions of magnetization and energy fluctuations.

The behavior of the local exponents suggests that the static exponents are universal when the FM density is between pure and multicritical values, and this universality class is different from those of pure and multicritical systems. The estimated values, $\nu = 0.67(2)$ and $\beta = 0.34(1)$, are consistent with the results from the Monte Carlo renormalization analysis [28]. Furthermore these values are consistent with the results for the diluted Ising system from renormalization group analysis [7], simulation [13] and experiment [8]. So the present results confirms the random-fixed-point picture for the cubic $\pm J$ Ising model. The estimated values of dynamical exponent z depend on p, although previous renormalization and Monte Carlo studies suggested dynamical universality [12,29]. It becomes larger when p is decreased.

At the multicritical point, the estimated values of exponents are $\nu = 0.90(2)$, $\beta = 0.42(2)$ and $z = 5.2(1)$. These values are consistent with the previous estimations from high-temperature expansion analysis [30].

The FM transition density along the Nishimori line is estimated to be $p_{\rm mc} = 0.8894(9)$, which is consistent with the conjecture [31],

$$p_{\rm mc} \log p_{\rm mc} + (1 - p_{\rm mc}) \log(1 - p_{\rm mc}) = -\frac{\log 2}{2}\,, \tag{6.6}$$

that is, $p_{\rm mc} = 0.889972\cdots$.

Acknowledgements

The vector-parallel computer Hitachi SR8000 of the of the Institute of Statistical Mathematics, and of the Institute of Solid State Physics of the University of Tokyo are used for the simulations. This work is partly supported by the Nestlé Science Promotion Committee and by the Japan Society for the Promotion of Science (No. 13740235).

References

1. B.M. McCoy and T.T. Wu: Phys. Rev. Lett. **21**, 549 (1968)
2. A.B. Harris: J. Phys. **C7**, 1671 (1974)
3. J.T. Chayes, L. Chayes, D.S. Fisher and T. Spencer: Phys. Rev. Lett. **57**,) 2999 (1986
4. A. Aharony, A.B. Harris and S. Wiseman: Phys. Rev. Lett. **81**, 2552 (1998)
5. A.B. Harris and T.C. Lubensky: Phys. Rev. Lett. **33**, 1540 (1974)
6. T.C. Lubensky: Phys. Rev. **B11**, 3537 (1975)
7. I.O. Mayer: J. Phys. **A22**, 2815 (1989)
8. T.R. Thurston, C.J. Peters, R.J. Birgeneau and P.M. Horn: Phys. Rev. **B37**, 9559 (1988)
9. D.P. Landau: Phys. Rev. **22**, 2450 (1980)
10. J. Marro, A. Labarta and J. Tejada: Phys. Rev. **B34**, 347 (1986)
11. H.O. Heuer: Phys. Rev. **B42**, 6476 (1990); J. Phys. **A26**, L333 (1993)
12. H.O. Heuer: J. Phys. **A26**, L341 (1993)
13. H.G. Ballesters, L.A. Fernandez, V. Martin-Mayor, A. Munoz Sudupe, G. Parisi and J.J. Ruiz-Lorenzo: Phys. Rev. **B58**, 2740 (1998)
14. K. Hukushima: J. Phys. Soc. Jpn. **69**, 631 (2000)
15. N. Ito: Physica A **192**, 604 (1993); **196**, 591 (1993)
16. N. Ito, T. Matsuhisa and H. Kitatani: J. Phys. Soc. Jpn. **67**, 1188 (1998)
17. N. Ito, K. Hukushima, K. Ogawa and Y. Ozeki: J. Phys. Soc. Jpn. **69**, 1931 (2000)
18. Y. Ozeki, N. Ito and K. Ogawa: *Activity Report 1999 of the Supercomputer Center of ISSP* (The University of Tokyo, 2000) p. 37
19. N. Ito and Y. Ozeki: *Computer Simulation Studies in Condensed Matter Physics XIII*, ed. by D.P. Landau, S.P. Lewis and H.-B. Schüttler (Springer, Berlin Heidelberg New York 2001) p. 175
20. N. Ito, S. Fukushima, H. Watanabe and Y. Ozeki, *Computer Simulation Studies in Condensed Matter Physics XIV*, ed. D. P. Landau, S. P. Lewis and H.-B. Schüttler (Springer-Verlag, Heidelberg, 2002) p.27.
21. Y. Ozeki and N. Ito: J. Phys **A31**, 5451 (1998)
22. N. Ito, Y. Ozeki and H. Kitatani: J. Phys. Soc. Jpn. **68**, 4547 (1999)
23. Y. Ozeki and N. Ito: Suppl. J. Phys. Soc. Jpn. **69**, 193 (2000)
24. N. Ito, K. Ogawa, K. Hukushima and Y. Ozeki: Prog. Theor. Phys. Suppl. **138**, 555 (2000)
25. Y. Ozeki and N. Ito: Phys. Rev. **B64**, 024416 (2001)
26. Y. Ozeki, N. Ito and K. Ogawa: J. Phys. Soc. Jpn. **70**, 3471 (2001)
27. H. Nishimori: Prog. Theor. Phys. **66**, 1169, (1981); **76**, 305 (1986); J. Phys. Soc. Jpn. **55**, 3305 (1986)
28. Y. Ozeki and H. Nishimori: J. Phys. Soc. Jpn. **56**, 1568 (1987); J. Phys. Soc. Jpn. **56**, 2992 (1987)
29. G. Grinstein, S.-K. Ma and G.F. Mazenko: Phys. Rev. **B15**, 258 (1977)
30. R.R.P. Singh: Phys. Rev. Lett. **67**, 899 (1991)
31. H. Nishimori and K. Nemoto: J. Phys. Soc. Jpn. **71**, 1198 (2002)

7 Nonequilibrium Relaxation Analysis of Frustrated XY Models in Two Dimensions

Y. Ozeki[1] and N. Ito[2]

[1] Department of Physics, Tokyo Institute of Technology, Oh-okayama, Meguro-ku, Tokyo 152-8551, Japan
[2] Department of Applied Physics, The Tokyo University, Hongo, Bunkyo-ku, Tokyo 113-8656, Japan

Abstract. We apply the nonequilibrium relaxation (NER) method to the fully-frustrated XY models on square and triangular lattices, in which the possibility of two different transitions, the chiral transition and the Kosterlitz–Thouless one, have been discussed. Further, we calculate the NER functions of fluctuations for these models, and analyze the universality class of them.

7.1 Introduction

The frustrated system is one of main subjects in statistical physics because of its rich phase diagrams and possibility of new universality classes [1]. The equilibrium Monte Carlo simulation (EMCS) has been widely used and revealed many helpful informations in phase transitions and critical phenomena. Although it works even in frustrated systems, it sometimes suffers difficulties in the analysis due to slow-relaxation in low temperature regime, which restrict the available system sizes too small.

The nonequilibrium relaxation (NER) method is an efficient numerical technique for analyzing equilibrium phase transitions [2–5]. One may observe the relaxation of the order parameter (e.g. the magnetization in the ferromagnet) in the thermalization process from the complete ordered state. It provides the critical temperature and critical exponents accurately [2,3]. The NER analysis is advantageous over the EMCS in two features: Firstly, for fixed time, the size dependence of NER function is exponentially small even at the critical point. So the value in the thermodynamic limit is easily estimated. Secondly, equilibration step is not necessary. Simulation is made only up to the steps when the asymptotic behavior indicates the equilibrium state. These advantage becomes more effective for slow-relaxation systems [4,5].

In this article, we apply the NER method to fully-frustrated (FF) XY models in two dimensions [1,6–9], the antiferromagnetic (AF) XY model on the triangular lattice and the FF XY model on the square lattice. In these models, the possibility of two different transitions, the chiral transition and the Kosterlitz–Thouless (KT) one, have been discussed [6,7]. Further, we calculate the NER functions of fluctuations for the models, and discuss the universality class of them.

7.2 Models

The Hamiltonian we consider is

$$\mathcal{H} = -J \sum_{\langle ij \rangle} \cos(\theta_i - \theta_j + A_{ij}) \, ,$$

where $J > 0$ and $0 \leq \theta_i \leq 2\pi$. The constant A_{ij} takes π for all bonds in the triangular system, which represents nothing but an AF interaction, and 0 for three of four bonds in a plaquette and π for the other one in the square system.

The ground state for the models has plaquettes with clockwise or counterclockwise rotation in a two subplaquette pattern. The angular difference is $2\pi/3$ for the triangular system and $\pi/4$ for the square system. The system has a Z_2 symmetry due to this clockwise-counterclockwise degeneracy and a continuous $U(1)$ symmetry due to the global rotational symmetry. Thus, in low temperature regime, two kinds of phases appear. One is the KT phase associated with the $U(1)$ symmetry in two dimensions and the chiral phase associated with the Z_2 symmetry.

Since the works in [6,7], there has been a controversy concerning to the transition temperatures, T_{KT} and T_{c}, for these transitions; the double transitions ($T_{\mathrm{KT}} < T_{\mathrm{c}}$) or the single transition ($T_{\mathrm{KT}} = T_{\mathrm{c}}$). Another controversy arose concerning to the universality class for the chiral transition [8]. Since the broken symmetry in the chiral transition is a discrete Z_2 one, which is the same as the Ising model, it is natural to expect that the universality class is in the 2D Ising class ($\nu = 1$, $\beta = 1/8 \cdots$). However, some numerical studies indicated different classes [9].

7.3 NER Analysis

In the NER analysis, the initial state of relaxation is chosen as one of ground states. That is three sublattice structure in the triangular system and four sublattice structure in the square system. Let us denote the initial value of spin as θ_i^0. In the present case, we calculate two kinds of order parameters, the spin orientational oder parameter

$$m(t) = \left\langle \frac{1}{N} \sum_i \cos[\theta_i(t) - \theta_i^0] \right\rangle$$

and the chiral order parameter

$$\kappa(t) = \left\langle \frac{1}{N} \sum_{(i \to j)} \sin[\theta_i(t) - \theta_j(t)] \right\rangle ,$$

where the summation for $(i \to j)$ is taken for all bonds with fixed directions and $\langle \cdots \rangle$ represents the dynamical averaging.

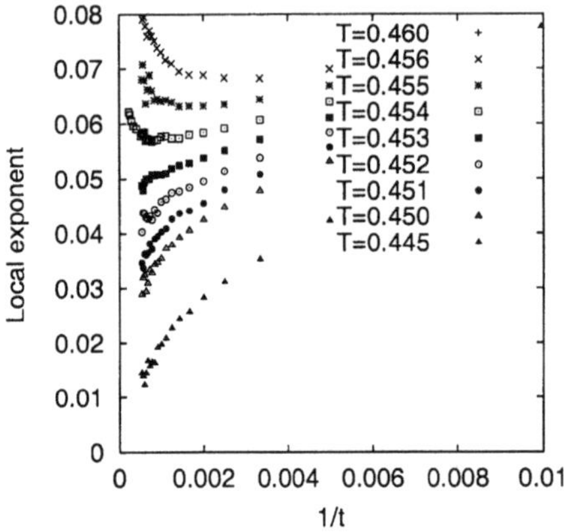

Fig. 7.1. Local exponent $\lambda(t)$ for the FF XY model on the square lattice

To estimate the chiral transition temperature, we analyze the relaxation of $\kappa(t)$ which decays in a power-law $\kappa(t) \sim t^{-\lambda_\kappa}$ at $T = T_c$, exponentially in the disordered regime and saturates to a finite value in the chiral phase. To distinguish the phases precisely, it is convenient to define the local exponent

$$\lambda(t) = -\frac{\partial \log \kappa(t)}{\partial \log t} \longrightarrow \begin{cases} \infty & (T > T_c) \\ \lambda_\kappa & (T = T_c) \\ 0 & (T < T_c) \end{cases} . \tag{7.1}$$

In Fig. 7.1, we plot $\lambda(t)$ for the square system. Calculations are carried out on 2000×2000 lattice with the skew boundary condition up to the observation time 5000 MCS. About $320 \sim 2800$ independent runs are performed for averaging. The size dependence is checked to be negligible, when we compare the data with those on 1500×1500 lattice for some temperatures. For efficiency of calculation, we discretize the spin state and use the 1200-states clock model. This discretization is also checked to be negligible, when we compare the data with the 800-state model. Hereafter we measure the temperature T in the unit of J/k_B. The bending up behavior indicating the disordered phase is observed in $T \geq 0.454$, and the bending down behavior indicating the ordered phase is observed in $T \leq 0.453$. Therefore, our estimation of the chiral transition point is $T_c = 0.4535(5)$ The similar analysis is made for the triangular system and we estimate $T_c = 0.512(1)$. They are almost consistent with those obtained so far.

Next, we analyze the universality class of the chiral transition. It is confirmed that the relaxation of fluctuations $f_{\kappa\kappa}(t) \equiv N\left[\langle\kappa(t)^2\rangle/\langle\kappa(t)\rangle^2 - 1\right]$ and $f_{\kappa e}(t) \equiv N\left[\langle\kappa(t)e(t)\rangle/\langle\kappa(t)\rangle\langle e(t)\rangle - 1\right]$, show power-law divergences at the critical point [3], where $e(t)$ is the energy per site. Assuming the dynamic-scaling hypothesis, one derives the relations for the asymptotic powers of relaxation functions with the standard critical exponents as $\lambda_\kappa = \beta/z\nu$, $\lambda_{\kappa\kappa} = d/z$ and $\lambda_{\kappa e} = 1/z\nu$. Thus one can estimate each exponent from a

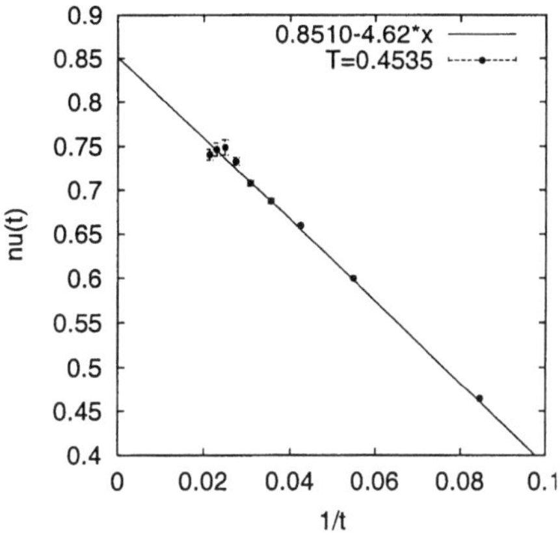

Fig. 7.2. Local exponent $\nu(t)$ for the FF XY model on the square lattice

combination of local exponents; e.g. $\nu(t) = \lambda_{\kappa\kappa}(t)/d\lambda_{\kappa e}(t)$. We plot $\nu(t)$ for the square system in Fig. 7.2, which indicates $\nu = 0.845(25)$. Calculations are made at the above estimated transition temperature on 500×500 square lattice up to the observation time 500 MCS. About 10^6 independent runs are performed for averaging. Similar analyses are made for other exponents and those in the triangular system (see Table 7.1). In the present accuracy, these results indicate that the universality class of the chiral transition in the 2D FF XY models is different from that in the 2D Ising model.

Finally, we estimate the KT transition temperature from the spin orientational order parameter $m(t)$. Unlikely to the chiral transition case, one can not evaluate the lower bound of T_{KT} since the relaxation is always power-law below T_{KT}. To estimate it, we propose a scaling relation [5] for $T > T_{\mathrm{KT}}$

$$m(t) = \tau^{\lambda} g(t/\tau) , \tag{7.2}$$

where τ is the relaxation time depending on the temperature, and λ is the dynamic exponent. Fitting the calculated curves to this relation, one obtains τ for each temperature. Then, estimate the value of T_{KT} using the relation

$$\tau(T) = A \exp(B/\sqrt{T - T_{\mathrm{KT}}}) , \tag{7.3}$$

which is expected from the similar relation for the correlation length in the KT system [10]. In Fig. 7.3, we plot the estimated relaxation time τ for the

Table 7.1. Summary of resulting transition temperatures and critical exponents

Lattice	T_{c}	T_{KT}	ν	β	η	z
Square	0.4535(5)	0.449(1)	0.851(25)	0.112(7)	0.28(3)	2.48(8)
Triangular	0.512(1)	0.508(1)	0.845(25)	0.105(8)	0.26(3)	2.51(7)

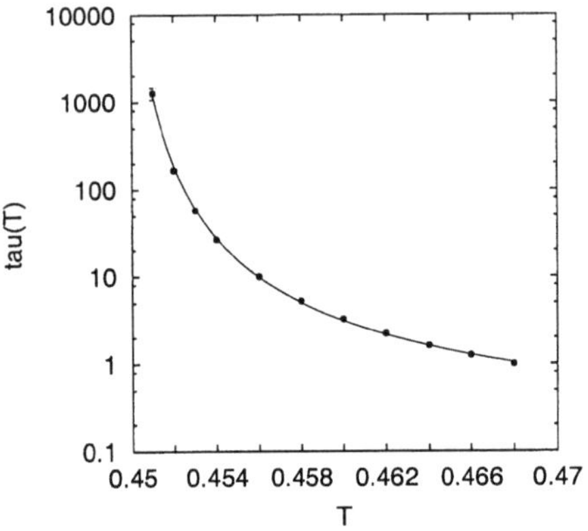

Fig. 7.3. Relaxation time for the FF XY model on the square lattice

square system and the fitting curve to (7.3) with $T_{\mathrm{KT}} = 0.449(1)$. Calculations are made on 1000×1000 square lattice up to the observation time 150000 MCS. About 320 independent runs are performed for averaging. The same analysis is made for the triangular system and obtain $T_{\mathrm{KT}} = 0.508(1)$. The results are summarized in Table 7.1. In both models, we have clearly observed the difference of T_{c} and T_{KT}, which indicates the double transitions and the existence of the intermediate phase.

7.4 Remarks

We have performed the NER analysis for the 2D FF XY models. The results confirm that the double transitions ($T_{\mathrm{c}} > T_{\mathrm{KT}}$) of chiral and KT phases occurs and the universality class of the chiral transition is different from that of the Ising model. To confirm the results more reliable, it is important to observe the longer time behavior (the finite time effect), while the simulated size is large enough to eliminate the finite size effect. Since the equilibration is not necessary in the NER analysis, the results does not suffer the difficulty due to the slow-relaxation in frustrated systems.

References

1. J. Villain: J. Phys. C**10** 4793 (1977); J. Phys. (Paris) **38** 26 (1977)
2. For a review, see Y. Ozeki, N. Ito and K. Ogawa: Activity Report 1999 (Supercomputer Center, ISSP Uinv. of Tokyo, 2000) and references therein.
3. N. Ito, K. Hukushima, K. Ogawa and Y. Ozeki: J. Phys. Soc. Jpn. **69** 1931 (2000)
4. K. Ogawa and Y. Ozeki: J. Phys. Soc. Jpn. **69** 2808 (2000)
5. Y. Ozeki, N. Ito and K. Ogawa: submitted to Phys. Rev. B
6. S. Teitel and C. Jayaprakash: Phys. Rev. B**27** 598 (1983)

7. S. Miyashita and H. Shiba: J. Phys. Soc. Jpn. **53** 1145 (1984)
8. D.H. Lee, J.D. Joannopoulos, J.W. Negele and D.P. Landau: Phys. Rev. Lett. **52** 433 (1984); Phys. Rev. B **33** 450 (1986)
9. See the recent articles and references therein, S. Lee and K.-C. Lee: Phys. Rev. B**49** 15184 (1994); B **57** 8472 (1998); P. Olsson: Phys. Rev. Lett. **75** 2758 (1995); Phys. Rev. B **55** 3585 (1997); G.S. Jeon, S.Y. Park and M.Y. Choi: Phys. Rev. B **55** 14088 (1997)
10. J.M. Kosterlitz: J. Phys. C **7** 1046 (1974)

8 First Principles Calculations of the Magnetic Structure in FeMn/Co Bilayers

B. Ujfalussy[1], T.C. Schulthess[2], and M. Stocks[2]

[1] University of Tennessee, Knoxville TN 08544, USA
[2] Oak Ridge National Laboratory
P.O. Box 2008-6114, Oak Ridge, TN 37831-6116, USA

Abstract. We perform a first principles spin-dynamics simulation of the $Fe_{0.5}Mn_{0.5}$/Co(111) interface in order to determine the orientational configuration of the magnetization in the ground state. We find that the magnetic configuration does not differ significantly from bulk Co in the Co, however it changes from a bulk 3Q configuration to a 1Q like antiferromagnetic configuration in $Fe_{0.5}Mn_{0.5}$. Additionally, the $Fe_{0.5}Mn_{0.5}$ moments are found to be mostly perpendicular to Co.

8.1 Introduction

Bilayer systems consisting of antiferromagnetic FeMn in contact with a ferromagnet (FM) have received considerable attention in recent years due to very favorable exchange bias that can be observed in these systems. Exchange bias refers to a shift of the magnetization curve away from the zero field axis, that is observed when the system is cooled in a field [1]. The peculiarity of the FeMn/FM system is that the exchange bias on set is at relatively high temperatures (taking advantage of the high Neel temperature of FeMn) and occurs without the large increase in coercivity, that is usually observed in exchange bias systems. However, despite these favorable magnetic properties, FeMn is very sensitive to oxidation and thus cannot be applied in device applications. Understanding the origin of the very favorable magnetic properties of FeMn/FM systems is thus crucial, in order to find suitable alternatives for FeMn as an antiferromagnet.

In this paper we are concerned with the details of the ground state magnetic structure in FeMn/FM systems. Experimental observations are suggestive of a perpendicular alignment of the FM moments with regard to AFM spins [2]. This type of magnetic structure, which originates from a spin-flop type of coupling at the interface, has been observed in numerous AFM/FM systems [3–5] and has been described theoretically within the context of a Heisenberg model [6–8]. Understanding perpendicular coupling in FeMn/FM is, however, far from straightforward for two reasons. (1) The bulk ground state structure of FeMn is a non-collinear 3Q state [9,10] that cannot be described in the context of a Heisenberg model. Applying the above mentioned theory of spin-flop type coupling to FeMn/FM is thus inadequate. (2) The 3Q has a threefold axis along the (111) direction and it is therefore unclear how one can align the FM perpendicular to the FeMn spins – note that this is different from the uniaxial AFM systems in which perpendicular coupling is usually observed.

In the present paper we will investigate the microscopic magnetic structure of FeMn/Co interfaces using first principles spindynamics simulations, originally pioneered by Antropov et al. (see [11]). Here Co is chosen instead of Permalloy to be the ferromagnet to simplify the description of the FM side of the bilayer. The results, can be expected to be quite generic and valid for most FeMn/FM systems.

8.2 Computational Details

When it is described in a frame of reference, where the z axis is along the (111) direction, perpendicular to the FeMn/Co interface, the fcc lattice can be described as an 18 atoms/unit cell entity. This unit cell has been repeated $4 \times 4 \times 7$ times to obtain the unit cell of the calculation. Such a system consist of 21 layers and 2016 atoms in total. The first 6 layers were decorated by Co atoms. In the initial state the Co is assumed ferromagnetic with the magnetic moment pointing in plane. The remaining 15 layers were randomly populated by Fe and Mn atoms, with magnetizatons nevertheles initialized according to the orderd "3Q" state, and preserving the 50%–50% concentration, respectivelly. This initial state for the most relevant interfacial layers are shown on Fig. 8.1.

Once the initial state has been set up, we used the method of first priciples spin-dynamics to study the evolution of this initial state into a ground state. A general orientational configuration of atomic magnetic moments $\{\boldsymbol{e}_i\}$ is not a ground state of LSDA. In order to properly describe the energetics of a general configuration it is necessary to introduce a constraint which maintains the orientational configuration. The purpose of the constraining fields is to force the local magnetization, $\boldsymbol{M}_i$, to point along the prescribed local direction $\boldsymbol{e}_i$. Note that at each site the constraining field ($\boldsymbol{B}_{\mathrm{cons}}$) is transverse to the magnetization. The constraining field is not known a priori, and has to be calculated by some additional algorithm. The details of the theory and the constrained local moment (CLM) model are contained in reference [12]. The CLM state may be considered as a time freeze during the evolution of the system from a general initial state to a final state. This final state may be the ground state or another non-equilibrium state. Currently we restricted the theory to include the former possibility only. We apply a simple model to describe the time-evolution process of the system between such CLM states: if the timestep is very small, the magnetic moments can be considered as classical moments of constant size. Relaxation of such moments

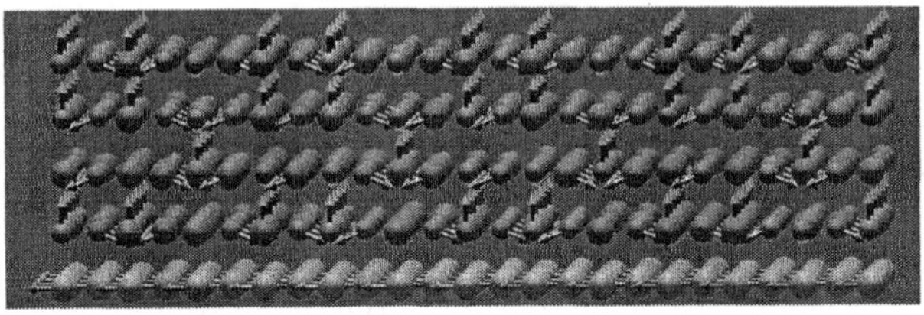

Fig. 8.1. Section of the FeMn/Co interface showing the initial configuration of the magnetic moments. The visualization shows the Co layer and four FeMn layers closest to the interface. The Co layer is initialized in a FM state and the FeMn layers are initialized in the 3Q structure

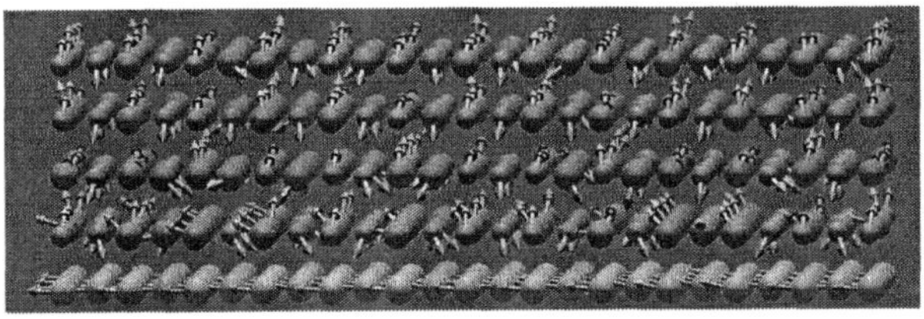

Fig. 8.2. Section of the FeMn/Co interface showing the final configuration of the magnetic moments for the same five layers as shown for the initial configuration

can be described by an over-damped Landau–Lifshitz equation of motion,

$$\begin{aligned} \frac{d}{dt}\boldsymbol{e}_i &= \alpha \boldsymbol{e}_i \times (\boldsymbol{e}_i \times \boldsymbol{B}_{\text{eff}}) \\ \boldsymbol{B}_{\text{eff}} &= -\boldsymbol{B}_{\text{cons}}\,, \end{aligned} \tag{8.1}$$

in which the precession term has been ommited since presently we are interested in relaxation to the ground state only. This simple first order differential equation needs to be solved for each moment to give the orientational configuration for the next timestep at which a new constrained LDA calculation is then performed. This is repeated until a ground state is obtained in which $\boldsymbol{B}_{\text{cons}} = 0$, so (8.1) leaves the system unchanged.

The LDA calculation in the presence of the constraining fields were performed using the LSMS method. The LSMS method [13] achieves $O(N)$ scaling by using real space multiple scattering theory to obtain the Green function and hence the electronic charge and magnetization densities. Therefore it is suitable for LDA calculations on very large systems, as was demonstrated previously. In the LSMS method the atoms comprising the crystal are treated as disjoint electron scatterers. The multiple scattering equations are solved for each atom by ignoring multiple scattering processes outside a local interaction zone (LIZ) centered on each atom which, in the case of the Co/FeMn interface was 87 atoms.

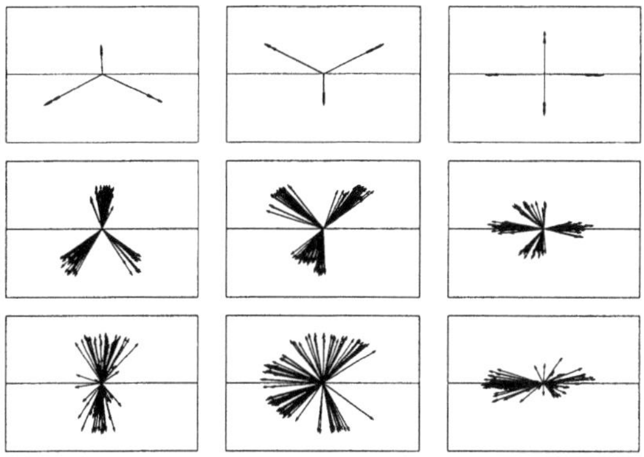

Fig. 8.3. Rose plot of the Fe/Mn layer moments adjacent to Co. The columns are the projections to the yz-, xz-, xy-planes, respectivelly. The various rows correspond to different timesteps, the first is the initial (3Q) state and the last is the final state

8.3 Results

Once the system was initialized into the state described above, we searched for the ground state using (8.1). The results for the most relevant secion of the final state are shown on Fig. 8.2. It is rather hard to visualize the results of this calculation in such a short paper. It is immediatelly apparent though that the moments of the Co atoms hardly move. In fact they do change about 3–4 degrees out of the surface on the top Co layer. The most dramatic change occures in the FeMn. It is easy to see, that the "3Q" state dissolves, and gives way to a new state. It is rather difficult to analize the nature of this new state, mostly because of the large amount of geometrical data due to the large sample size. Also, because FeMn is chemically a solid solution, it necessarily contains a lot of fluctuations. It is important to mention that according to our calculations [10], even in bulk $Fe_{0.5}Mn_{0.5}$ alloys the magnetization deviates from the ideal "3Q" state and fluctuates around the ideal "3Q" directions a few degrees. This is the consequence of the microscopically different local environments. However, to gain some insight into this new state, in Fig. 8.3 we show a rose plot of three timesteps during the calculation. It can be easily seen that the 3Q state dissolves, and reorders into something which resembels an antiferromagnetic state perpendicular to the Co momentums. This more conventional antiferromagnetically coupled system appears to be essentially a "1Q" state which is additionally perpendicular to the Co. Of course this is true only on the average, as substantial fluctuations from the average do occur. The magnitude of the Co moments remain relatively unchanged, fluctuating around $\approx 1.6\,\mu_B$. Similarly, the Mn moments also appear to be remarkably unaffected, however the Fe moments fluctuate quite a bit, between, $1.5\,\mu_B$ and $2\,\mu_B$.

8.4 Discussion

The main result of these calculations is that the structure of the magnetic ground state of FeMn changes from a non-collinear 3Q structure to a 1Q structure when FeMn is in proximity to ferromagnetic Co. In agreement with experimental observations, the FM moments align perpendicular to the AFM moments, which now is in a quasi-uniaxial structure. Despite the large size of the unit cell used in our calculation (large on the scale of first principles all electron calculations), the systems considered here was still very small compared to realistic thicknesses of the FeMn layer. Calculations with thicker FeMn layers using the layered KKR techniques are presently underway.

Acknowledgements

Calculations were performed at CCS-ORNL and at NERSC. This work was supported by the DOE-OS through BES-DMSE and OASCR-MICS under contract number DE-AC05-00OR22725 with UT-Battelle LLC.

References

1. For details on exchange bias and related effects see the recent reviews, J. Nogués and I.K. Schuller: J. Magn. Magn. Mater. **192**, 203 (1999); A.E. Berkowitz and K. Takano: J. Magn. Magn. Mater. **200**, 552 (1999); M. Kiwi, J. Magn. Magn. Matter. **234**, 584 (2001); and references theirin
2. R. Junblut, R. Coehoorn, M.T. Johnson, C. Sauer, P.J. van der Zaag, A.R. Ball, T.G.S.M. Rijks, J. aan de Steege, and A. Reinders: J. Magn. Magn. Mater. **148**, 300 (1995)
3. T.J. Moran and I.K. Schuller: J. Appl. Phys. **79**, 5109 (1996)
4. Y. Ijiri, J.A. Borchers, R.W. Erwin, S.-H. Lee, P.J. van der Zaag, and R.M. Wolf: Phys. Rev. Lett. **80**, 608 (1998)
5. T.J. Moran, J. Nogués, D. Lederman, and I.K. Schuller: Appl. Phys. Lett. **72**, 617 (1998)
6. L.L. Hinchey and D.L. Mills: Phys. Rev. B **34**, 1689 (1986)
7. N.C. Koon: Phys. Rev. Lett. **78**, 4865 (1997)
8. T.C. Schulthess and W.H. Butler: Phys. Rev. Lett. **81**, 4516 (1998); J. Appl. Phys. **85**, 5510 (1999)
9. H. Umebajashi and Y. Ishikawa: J. Phys. Soc. Japan **21**, 1281 (1966); S.J. Kennedy and T.J. Hicks: J. Phys. F **17**, 1599 (1987)
10. T.C. Schulthess, W.H. Butler, G.M. Stocks, S. Maat, G.J. Mankey: J. Appl. Phys. **85**, 4842 (1999); G.M. Stocks, B. Ujfalussy, W.A. Shelton, T.C. Schulthess, W.H. Butler, A. Canning: J. Appl. Phys. (in press)
11. V.P. Antropov et al.: Phys. Rev. **54**, 1019 (1996)
12. B. Ujfalussy, X.D Wang, D.M.C. Nicholson, W.A. Shelton, G.M. Stocks, Y. Wang, B.L. Gyorffy: J. Appl. Phys. 85, 4824 (1999)
13. Y. Wang, G.M. Stocks, W.A. Shelton and D.M.C. Nicholson: Phy. Rev. Letters **75**, 2867 (1995)

Part II

Quantum Systems

9 Phase Transitions and Quantum Effects in Pore Condensates

J. Hoffmann[1] and P. Nielaba[2]

[1] Fraunhofer-Institut für Techno- und Wirtschaftsmathematik, Gottlieb-Daimler-Str. 49, 67663 Kaiserslautern, Germany
[2] Physics Department (Theory), University of Konstanz, 78457 Konstanz, Germany

Abstract. We investigate structural properties of pore condensates in cylindrical pores as well as phase transition and quantum effects by path integral Monte Carlo simulations.

9.1 Introduction

Nanostructures in reduced geometry have become an interesting research domain in the last years. Despite the fact that by experimental techniques many structural-, elastic-, electronic-,and phase- properties of systems in the size of a few nanometers have been obtained, the theoretical investigations and analyses are still in an initial stage. This is partly due to the fact that systems which are far away from the thermodynamic limit (with infinitely many particles) due to their finite size are difficult to handle by analytical methods which are suitable for systems with either few particles (2–5) or in the limit of infinitely many particles. In this field computer simulations have become more and more important since nano-systems in reduced geometry contain about 10–10.000 particles, which is nearly ideal for the application of computer simulation methods. Here we investigate structural properties of pore condensates in cylindrical pores as well as phase transition and quantum effects by path integral Monte Carlo simulations.

9.2 The Method

Many of the interesting effects in pore condensates take place at low temperatures so that the consideration of quantum mechanics is important. In order to quantify quantum effects we have performed path integral Monte Carlo (PIMC) simulations.

Canonical averages $< A >$ of an observable A in a system defined by the Hamiltonian $\mathcal{H} = E_{\rm kin} + V_{\rm pot}$ of N particles in a volume V are given by:

$$\langle A \rangle = Z^{-1} \mathrm{tr}[A \exp(-\beta \mathcal{H})] \,. \tag{9.1}$$

Here $Z = \mathrm{tr}[\exp(-\beta\mathcal{H})]$ is the partition function and $\beta = 1/k_{\mathrm{B}}T$ is the inverse temperature. Utilizing the Trotter-product formula,

$$\exp(\beta\mathcal{H}) = \lim_{P\to\infty} (\exp(-\beta E_{\mathrm{kin}}/P)\exp(-\beta V_{\mathrm{pot}}/P))^P , \tag{9.2}$$

we obtain the path integral expression for the partition function:

$$Z(N,V,T) = \lim_{P\to\infty} \left(\frac{mP}{2\pi\beta\hbar^2}\right)^{3NP/2} \prod_{s=1}^{P} \int d\{\boldsymbol{r}^{(s)}\} \times\exp\left\{ -\frac{\beta}{P}\left[\sum_{k=1}^{N} \frac{mP^2}{2\hbar^2\beta^2}(\boldsymbol{r}_k^{(s)} - \boldsymbol{r}_k^{(s+1)})^2 + V_{\mathrm{pot}}(\{\boldsymbol{r}^{(s)}\})\right]\right\} . \tag{9.3}$$

Here, m is the particle mass, integer P is the Trotter number and $\boldsymbol{r}_k^{(s)}$ denotes the coordinate of particle k at Trotter-index s, and periodic boundary conditions apply, $P+1=1$. This formulation of the partition function allows us to perform Monte Carlo simulations [1] for increasing values of P approaching the true quantum limit for $P \to \infty$.

Thermal averages in the ensemble with constant pressure p are given via the corresponding partition function $\Delta(N,p,T) = \int_0^\infty dV^{-\beta pV} Z(N,V,T)$. In (9.1) we see that in the path integral formalism each quantum particle k can (for finite P-values) be represented by closed quantum chains of length P in position space where the classical coordinate of the point $\boldsymbol{r}_k^{(s)}$ on this chain at the Trotter index s has a harmonic interaction to its nearest neighbors at $\boldsymbol{r}_k^{(s+1)}$ and $\boldsymbol{r}_k^{(s-1)}$. An interaction between different quantum particles takes places only between particles $\{\boldsymbol{r}^{(s)}\}$ with the same Trotter index s. Due to this property the entire system with Trotter index s can be placed efficiently in one processor of a parallel computer with P processors, where the potential energy of all N particles can be computed for this Trotter index s (with an effort $\propto N(N-1)/2$).

9.3 Phase Transitions in Nanopores

Phase transitions of pore condensates in nano-pores (i.e. Vycor) have been investigated by experimental methods recently [2,3]. Besides spinodal decomposition, phase transition temperature reductions have been studied for cylindrical nano-pores with small diameters.

With computer simulations we have analyzed [4,5] many interesting properties of "Ar"- and "Ne"- pore condensates recently [modeled as Lennard–Jones systems with particle diameter σ and interaction energy ϵ, in our computations we use particle masses $\mathrm{m}^* = \mathrm{m}\sigma^2\varepsilon/\hbar^2 = 100$ and $\mathrm{m}^* = 1000$ for simplicity well approximating the particle masses of Ne and Ar ($\mathrm{m}^* = 112$ and $\mathrm{m}^* = 1160$)].

The effective wall potential has been computed by integration over the matrix, modeled as a Lennard–Jones-continuum, Fig. 9.1 shows a schematic

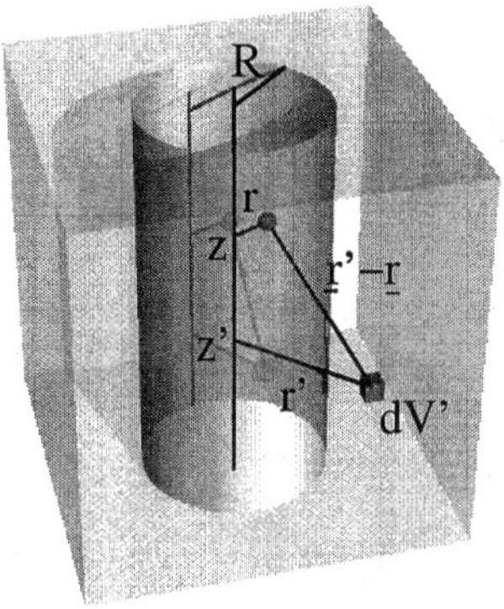

Fig. 9.1. Schematic picture of the system geometry. In the computation of the wall potential for a particle inside the pore at position $\boldsymbol{r}$ an (continuum-) integration over Lennard–Jones "particles" at position $\boldsymbol{r}'$ outside the pore has been done

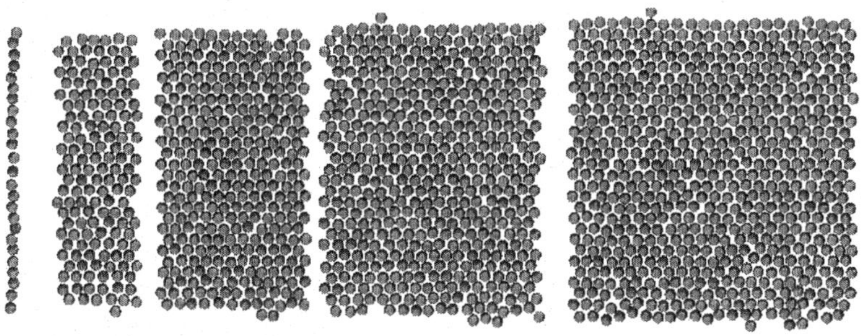

Fig. 9.2. Radial-layers-resolved (unrolled) pore Ar-condensate in cylindrical pore with diameter $5\,\sigma$ ($T^* = 0.34$)

picture of the system geometry. These systems have – like the gas-liquid-transition in "bulk"-systems- an adsorbate-condensate phase transition at low temperatures, the precise shape of the phase diagram is strongly influenced by the system geometry (pore radius). In turns out that with increasing attractive wall interaction the critical density increases, the adsorbate density increases strongly, and the condensate density increases weakly. A meniscus is formed with increasing curvature, the configurations become less stable and the critical temperature decreases.

The critical temperature is reduced with decreasing pore diameter. Beginning from the wall a formation of layered shell structures is found which may favor or disfavor the occupancy of sites at the pore axis due to packing effects. The density at the pore axis thus is an oscillatory function of the pore radius with density maxima for pore radii of $n\sigma$ and minima for pore radii of $(n + 1/2)\sigma$.

For large pore diameters the density oscillations decay from the wall towards the pore axis and the system approaches the "bulk".

At lower temperatures we obtain [4,5] a phase transition into a solid phase with long ranged positional order. In this solid phase cylindrical shells are formed with triangular lattice structures in azimuthal direction. In Fig. 9.2

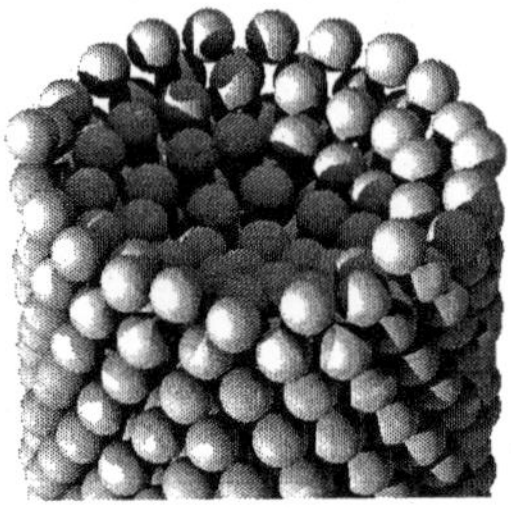

Fig. 9.3. Typical condensate configuration ($T^* = 0.31$) showing the meniscus structure

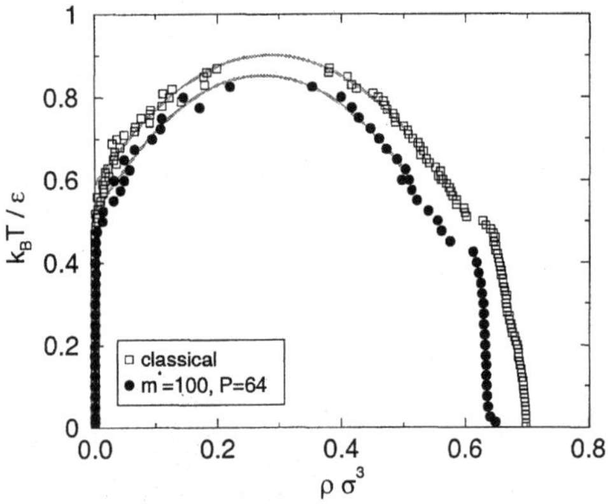

Fig. 9.4. Phase diagrams of (classical) Ar and Ne in cylindrical pores (radius $3.5\,\sigma$, length $60\,\sigma$, N = 650). Comparison of PIMC and classical MC. Lines: quadratical fits through the data ($T^* > 0.6$) for $m^* = 100$, P = 64 and the classical case

we show such unrolled layers for a pore with radius 5σ in a NVT-ensemble simulation indicating a meniscus shaped interface in the solid phase (see Fig. 9.3).

The meniscus curvature decreases with the temperature in the solid as well as in the fluid phase. The triple point temperature is influenced by the geometrical finite size effects (pore radius) as well as by the wall-particle interaction. Increasing interaction strength results in an increasing freezing temperature. Only a small radius effect on the triple point temperature was detected.

The geometrical finite size effect of the pore radius results in a packing effect with preferential occupancy of sites at the pore axis for pores with diameters $3.75\,\sigma$ and $4.75\,\sigma$, whereas for diameters of $3.25\,\sigma$ and $4.25\,\sigma$ the density at the axis is reduced. Beyond these results a two step phase transition from the fluid to the solid phase was found in agreement with results obtained in experimental studies of specific heat capacities at the freezing of Ar in Vycor [6].

By path integral Monte Carlo simulations [7–10] the effect of the quantum mechanics on the potential energy as a function of the temperature has been quantified [4,5]. In contrast to classical simulations we obtain by PIMC simulations for Ar- and Ne-condensates an horizontal temperature dependency of the energy resulting in a decrease of the specific heat to zero at small temperatures in agreement with the third law of thermodynamics. The resulting phase diagram for Ar- and Ne-condensates and a comparison with classical computations is shown in Fig. 9.4. In the Ne-system (containing the lighter particles) a significant reduction (by about 5–10%) of the critical temperature is found due to quantum delocalizations as well as a strong reduction of the solid density and a crystal structure modification in comparison with the classical case.

Corresponding PIMC studies for N_2- and CO-pore condensates [4,5] reveal an interesting and rich phase behavior resulting from the interplay of rotational and translational degrees of freedom.

Acknowledgments

We thank the HLRS and the HLRZ for computer time. This work was supported by the SFB 513 and the DFG (Ni 259/8-2).

References

1. D.P. Landau, K. Binder: *A Guide to Monte Carlo Simulations in Statistical Physics* (Cambridge University Press 2000)
2. Z. Zhang, A. Chakrabarti: Phys. Rev. E **50**, R4290 (1994); J.C. Lee: Phys. Rev. Lett. **70**, 3599 (1993); A. Chakrabarti: Phys. Rev. Lett. **69**, 1548 (1992); A.J. Liu, G.S. Grest: Phys. Rev. A **44**, R7894 (1991); A.J. Liu, D.J. Durian, E. Herbolzheimer, S.A. Safran: Phys. Rev. Lett. **65**, 1897 (1990); L. Monette, A.J. Liu, G.S. Grest: Phys. Rev. A **46**, 7664 (1992); Z. Zhang, A. Chakrabarti: Phys. Rev. E **52**, 2736 (1995)
3. M.W. Maddox, K.E. Gubbins, M. Sliwinska-Bartkowiak, S.-H. Suh: Mol. Simulat. **17**, 333 (1997); M.W. Maddox, K.E. Gubbins: J. Chem. Phys. **107**, 9659 (1997); R. Radhakrishnan, K.E. Gubbins: Phys. Rev. Lett. **79**, 2847 (1997); L.D. Gelb, K.E. Gubbins: Phys. Rev. **E 56**, 3185 (1997); P. Huber, K. Knorr: Phys. Rev.**B 60**, 12657 (1999)
4. J. Hoffmann: Ph.D.-Thesis, University of Konstanz (2002)
5. J. Hoffmann, P. Nielaba, in "High Perfomance Computing in Science and Engineering '01", ed. by E. Krause and W. Jäger (Springer, Berlin Heidelberg New York 2002) pp. 92; J. Hoffmann, P. Nielaba, preprint
6. D. Wallacher, K. Knorr: Phys. Rev. **B63**, 104202 (2001)
7. P. Nielaba, in: *Computational Methods in Surface and Colloid Science*, M. Borowko (Ed.) (Marcel Dekker, New York 2000) pp. 77–134
8. M. Presber, D. Löding, R. Martonak, P. Nielaba: Phys. Rev. B **58**, 11937 (1998)
9. M. Reber, D. Löding, M. Presber, Chr. Rickward, P. Nielaba: Comp. Phys. Commun. **121–122**, 524 (1999)
10. C. Rickwardt, P. Nielaba, M.H. Müser, K. Binder: Phys. Rev. **B63**, 045204 (2001)

10 Polynomial Expansion Method for the Monte Carlo Calculation of Strongly Correlated Electron Systems

N. Furukawa[1] and Y. Motome[2]

[1] Department of Physics, Aoyama Gakuin University, Setagaya, Tokyo 157-8572, Japan
[2] Tokura SSS Project, ERATO, Japan Science and Technology Corporation, c/o AIST, Tsukuba 305-8562, Japan

Abstract. We present a new Monte Carlo algorithm for a class of strongly correlated electron systems where electrons are strongly coupled to thermodynamically fluctuating classical fields. As an example, the method can be applied to electron-spin coupled systems such as the double-exchange model and dilute magnetic semiconductors, as well as electron-phonon systems. In these systems, calculation of the Boltzmann weights involves quantum-mechanical treatments for the electronic degrees of freedom. The traditional method requires all eigenvalues of the electronic Hamiltonian through a matrix diagonalization. Here we demonstrate that the Boltzmann weight can be obtained by matrix products and trace operations, if the calculation of the density of states is performed through a moment expansion using the Chebyshev polynomials of the Hamiltonian matrix. Therefore, the polynomial expansion method provides a faster calculation for the Boltzmann weights on large-size systems. We also show some results for the application of this algorithm to the study of critical phenomena in the simplified double-exchange model.

10.1 Introduction

Physics of strongly correlated electron systems is one of the most attractive fields in the studies of condensed matter phenomena [1]. Since the discovery of the high-T_c cuprates, various electron-electron interacting systems such as the Hubbard model as well as the t–J model have been re-investigated extensively. Due to the inherent difficulties in the models known as the many-body problems, however, there remain unexplored physics in these systems.

There exists another class of strongly interacting systems, where conduction electrons are strongly coupled to some external field degrees of freedom. One of the models categorized to this class is the double-exchange model which has been introduced to explain the ferromagnetism as well as colossal magnetoresistance of perovskite manganites [2]. In this model, electrons are strongly coupled to localized spins through Hund's-rule couplings. An electron–phonon system is another example. In addition, models with electron–electron interactions can in general be mapped to those where electrons are coupled to auxiliary fields by Stratonovich–Hubbard transformations. Therefore, this class involves most of the interesting electronic systems.

Let us hereafter concentrate on models in which the external field variables are treated as classical fields. Typical examples are the electron-lattice coupled systems where lattice distortions are treated as adiabatic fields, as well as the double-exchange model if localized spins are treated as classical rotators. Models with quantum fields can be approximated by classical fields if the fields cause finite temperature phase transitions. Near the critical point, correlation length of the quantum fields in the imaginary-time axis direction becomes longer than the inverse temperature, so that the field can be viewed as static. This is known as the renormalized classical regime.

The problem of electrons coupled to classical fields seems simple at the ground state. At finite temperatures, however, and especially near the critical points, it becomes highly nontrivial if the couplings between electrons and fields are strong. Dynamics of electrons is affected by randomly-distributing potentials whose variations might be much stronger than the kinetic energy. The fields have thermal fluctuations which might be nearly diverging.

Let us emphasize that classical or adiabatic fields affect electronic states in notable ways. If the fields are dynamical, with a typical timescale which is faster than that of the electron motion, the electrons feel the average of the potential fields. Then, the electrons are not influenced by the fluctuations. However, if the dynamics of the fields are slow, or the fields are static, each electron can accurately follow the field fluctuations. Then, the electronic states are strongly affected by the fluctuations.

In such cases, there occur breakdowns of the "band picture" which is based on a translational symmetry. Fluctuating fields cannot be treated by Hartree–Fock type mean-field treatments. Perturbational approaches are not applicable for the cases of strong coupling. A promising approach to investigate thermodynamics of these models is by numerical calculations, which will be shown in the following sections.

10.2 Thermodynamics of Electronic Systems Coupled with Classical Fields

10.2.1 Model Hamiltonian

Examples. Here we introduce Hamiltonians for various electronic systems which are coupled with classical fields. A typical candidate which belongs to such a class is the electron system coupled to adiabatic lattice distortions. As an example, a Hamiltonian for tight-binding spinless electrons coupled to local lattice distortions $\{\phi_i\}$ may be described in a form

$$\hat{\mathcal{H}} = -t \sum_{<ij>} (c_i^\dagger c_j + h.c.) + g \sum_i c_i^\dagger c_i \cdot \phi_i + \frac{K}{2} \sum_i \phi_i^2 , \tag{10.1}$$

where $\{\phi_i\}$ are treated as adiabatic fields. Here, c_i ($c_i^\dagger$) is a spinless electron annihilation (creation) operator, while t and K represent an electron hopping energy and a lattice elastic constant, respectively. In this case, the lattice distortion field $\{\phi_i\}$ modulates local electron potentials through an electron–phonon coupling constant g. Another example which belongs to this category is a Peierls system in a form

$$\hat{\mathcal{H}} = -t \sum_{<ij>} (1 + g\phi_{ij})(c_i^\dagger c_j + h.c.) + \frac{K}{2} \sum_{<ij>} \phi_{ij}^2 , \tag{10.2}$$

where ϕ_{ij} is the bond distortion between adjacent i-th and j-th site. Here, bond distortion fields modulate electron hoppings.

Some strongly correlated models which are introduced to study magnetisms and charge orderings are also be considered. An example is the double-exchange model. The Hamiltonian is given in the form

$$\hat{\mathcal{H}} = -t \sum_{<ij>,\sigma} (c_{i\sigma}^\dagger c_{j\sigma} + h.c.) - J_{\mathrm{H}} \sum_i \boldsymbol{\sigma}_i \cdot \boldsymbol{S}_i . \tag{10.3}$$

Here, $\boldsymbol{\sigma}_i$ is the spin operator for conduction electrons defined by

$$\sigma_i^\alpha = \sum_{s,s'} c_{is}^\dagger \sigma_{ss'}^\alpha c_{is'} , \quad \alpha = (x, y, z) , \tag{10.4}$$

where $\sigma_{ss'}^\alpha$ is the Pauli matrix. Local spins $\boldsymbol{S}_i$ create effective magnetic fields coupled to the conduction electrons through the Hund's-rule coupling J_{H}, which involve both diagonal and off-diagonal terms with respect to spin indices of the electrons. Random depletion of local spins leads to a model for dilute magnetic semiconductors. The Falicov–Kimball model,

$$\hat{\mathcal{H}} = -t \sum_{<ij>} (c_i^\dagger c_j + h.c.) + U \sum_i c_i^\dagger c_i \cdot n_i', \quad n_i' = 0 \text{ or } 1 , \tag{10.5}$$

can also be categorized in this family. Here, electrons are coupled to static charge-density fields $\{n_i'\}$ through Coulomb repulsions denoted by U.

General Forms. In general, we consider a system described by

$$\hat{\mathcal{H}}(\{x_r\}) = \hat{\mathcal{H}}^{\mathrm{e}}(\{x_r\}) + V(\{x_r\}) , \tag{10.6}$$

where $\hat{\mathcal{H}}^{\mathrm{e}}$ is the electron-field coupled part while $V(\{x_r\})$ is the additional potential energy for the field part expressed by c-number. The electron-field part consists of terms such as electron hoppings and one-body potential energies, which are expressed by quadratic terms of the electron operators as

$$\hat{\mathcal{H}}^{\mathrm{e}}(\{x_r\}) = \sum_{ij} c_i^\dagger H_{ij}^{\mathrm{e}}(\{x_r\}) c_j . \tag{10.7}$$

In this notation, i denotes a combination of site, spin, and orbital indices, if any. Many body electron–electron interactions which involve more than quartic terms of electron operators, as well as interactions between electrons and some quantum fields are out of the scope of this scheme. As is often for the case of models which represent realistic materials, we assume short-range electron hoppings. Then, the Hamiltonian matrix H^{e} becomes a sparse matrix.

We consider a system on a finite size lattice, and the number of sites is defined by N hereafter. The dimension of the matrix H^{e} is proportional to N, as is clear from the definition in (10.7). An amount of a memory space necessary for a computation is at most $O(N^2)$. The matrix H^{e} depends on field variables $\{x_r\}$ which are defined on sites or bonds described by r. The number of degrees of freedom for $\{x_r\}$ scales as $O(N)$.

10.2.2 Thermodynamics

We first consider a moment when $\{x_r\}$ are fixed to some random values. Then, the Hamiltonian expresses a one-body electron system coupled with "random fields". Eigenenergies and wavefunctions can be obtained by diagonalizing the matrix H^{e}. In order to discuss the thermodynamics of the system at finite temperature, however, we need to consider thermal fluctuations of the field $\{x_r\}$ and take appropriate thermodynamic averages.

Partition Function. The fluctuation should be thermodynamically determined by $\hat{\mathcal{H}}$ through the partition function, which is written as

$$Z = \mathrm{Tr}_{\mathrm{C}}\mathrm{Tr}_{\mathrm{F}} \exp\left(-\beta\left[\hat{\mathcal{H}}^{\mathrm{e}}\left(\{x_r\}\right) - \mu\hat{N}_{\mathrm{e}}\right]\right) \cdot \exp\left(-\beta V(\{x_r\})\right) . \quad (10.8)$$

Tr_{C} and Tr_{F} are traces over classical degrees of freedom denoted by classical variables $\{x_r\}$ and fermion degrees of freedom, respectively. Here, β is an inverse temperature, μ is a chemical potential and $\hat{N}_{\mathrm{e}}$ is a particle-number operator.

A grand canonical trace over fermion degrees of freedom is easily calculated through eigenvalues of the Hamiltonian matrix as

$$\mathrm{Tr}_{\mathrm{F}} \exp\left(-\beta\left[\hat{\mathcal{H}}^{\mathrm{e}}\left(\{x_r\}\right) - \mu\hat{N}_{\mathrm{e}}\right]\right) = \prod_{\nu}\left\{1 + \exp\left[-\beta\left(\varepsilon_\nu - \mu\right)\right]\right\} . \quad (10.9)$$

Here, the product should be performed over all the eigenvalues ε_ν which also depend on field variables $\{x_r\}$ in an implicit way. Then, the system can be mapped to a classical system where the partition function can be written as

$$Z = \mathrm{Tr}_{\mathrm{C}} \exp\left[-S_{\mathrm{eff}}\left(\{x_r\}\right) - \beta V(\{x_r\})\right] , \quad (10.10)$$

with the effective action

$$S_{\mathrm{eff}}\left(\{x_r\}\right) = -\sum_{\nu} \log\left[1 + \exp\left(-\beta\left(\varepsilon_\nu - \mu\right)\right)\right] . \quad (10.11)$$

Probability distribution of a configuration of $\{x_r\}$ is given by

$$P(\{x_r\}) \propto \exp\left[-S_{\mathrm{eff}}\left(\{x_r\}\right) - \beta V(\{x_r\})\right] . \quad (10.12)$$

Thermodynamic quantities are obtained by taking averages of corresponding quantities over configurations $\{x_r\}$ through $P(\{x_r\})$.

We see that effective interactions between fields $\{x_r\}$ are determined by the quantum mechanics of the conduction electrons. Modulation of $\{x_r\}$ creates "random" potential fields to the conduction electrons, so that they cannot be viewed as band electrons, especially in the case of strong electron-field interaction systems. The electron eigenenergies in (10.11) are affected by $\{x_r\}$ in a complex way which cannot be described by perturbation theories. The functional form of $S_{\text{eff}}(\{x_r\})$ is highly nontrivial, and may not be described in an simple analytic form.

Monte Carlo Calculation. Since the trace in (10.10) is a multi-dimensional integration/summation, Monte Carlo (MC) method provides us an efficient way to calculate the thermodynamics of the system. A MC update is performed using the Boltzmann weight of the configuration $\{x_r\}$ which is given by $P(\{x_r\})$ in (10.12). Configurations of classical variables $\{x_r\}$ are updated by, e.g., the Metropolis algorithm. In this case, the ratio of the Boltzmann weights for current variables $\{x_r\}$ and trial variables $\{x'_r\}$ which is defined by $r = P(\{x'_r\})/P(\{x_r\})$ determines the probability for accepting the trial variables in the form $p = \min(r, 1)$. Note that the Boltzmann weight (10.12) is positive definite. We do not have a negative sign problem in this fermionic MC calculation.

10.2.3 Diagonalization Method for the Calculation of the Boltzmann Weight

A straightforward way to evaluate the effective action (10.11) is to calculate eigenenergies $\{\varepsilon_\nu\}$ through a diagonalization of the one-body Hamiltonian matrix H^{e}. Since the matrix dimension is $O(N)$, it is not difficult to diagonalize the Hamiltonian matrix as long as the memory size is concerned, compared to the case of many-body systems which involve electron–electron interactions or electrons coupled to bosonic fields where the memory size diverges exponentially.

However, there exist some difficulties in performing the MC calculations using the diagonalization method. Diagonalization of a matrix requires a computational complexity of $O(N^3)$. If the MC updates for $\{x_r\}$ are implemented by local update algorithms, the total complexity for one MC updates scales as $O(N^4)$ since the number of fields is proportional to N. It is needless to say that the calculation is repeated N_{step} times, where N_{step} is the MC steps which is typically taken as $10^3 \sim 10^6$. Although it is effortless to perform a calculation on a small cluster, a systematic analysis for increasing system sizes is quite difficult. Especially, for electronic systems, it is necessary to fix the electron density while increasing the cluster size, in order to make appropriate estimates for the thermodynamic limit with accuracy. This restricts us for the choice of the sizes and shapes of clusters. An example for such a restriction which we often encounter at three-dimensional electron systems is that the system size should be taken as $N = L \times L \times L$ with $L = 4, 6, 8, \ldots$. In this case, CPU time scales as L^{12}.

Therefore, an alternative method to calculate clusters with considerably large system sizes within reasonable CPU timescales and accuracies is desired.

10.3 Polynomial Expansion Monte Carlo Method

10.3.1 Polynomial Expansion Method for the Calculation of the Boltzmann Weight

Now we introduce the Chebyshev polynomial expansion method in order to accelerate the calculation of the effective action $S_{\text{eff}}(\{x_r\})$ and the Boltzmann weight $P(\{x_r\})$ [3]. The method is originally introduced to calculate density of states of large-size matrices [4,5]. In ordered to make this article self-contained, we start from general properties of the Chebyshev polynomials, followed by an algorithm for the calculation of the effective action $S_{\text{eff}}(\{x_r\})$ using the moment expansion.

Chebyshev Polynomials. The Chebyshev polynomials $T_m(x)$ are defined on the region $-1 \leq x \leq 1$ using the recursion formula

$$T_{m+1}(x) = 2xT_m(x) - T_{m-1}(x), \tag{10.13}$$

with $T_0(x) = 1$ and $T_1(x) = x$. We see that $T_m(x)$ are odd-(even-)functions for odd (even) number of m. Alternatively, they can be defined by $T_m(x) = \cos(m \cos^{-1}(x))$, which leads to an addition theorem

$$T_{m+m'}(x) + T_{m-m'}(x) = 2T_m(x)T_{m'}(x). \tag{10.14}$$

Especially, from (10.14), we have

$$T_{2m}(x) = 2T_m{}^2(x) - T_0(x), \tag{10.15a}$$

$$T_{2m+1}(x) = 2T_m(x)T_{m+1}(x) - T_1(x). \tag{10.15b}$$

There exists the orthonormal property in the form

$$\int_{-1}^{1} \frac{\mathrm{d}x}{\pi\sqrt{1-x^2}} T_n(x)T_m(x) = \begin{cases} 0, & m \neq n \\ 1, & m = n = 0 \\ 1/2, & m = n \neq 0 \end{cases}. \tag{10.16}$$

This allows us to make a series expansion of a generic function using the Chebyshev polynomials.

Moment Expansion of the Effective Action S_{eff}. We apply these properties to calculate the effective action S_{eff}. In the present case, (10.11) is rewritten as

$$S_{\text{eff}} = -\int \mathrm{d}\varepsilon D(\varepsilon) s(\varepsilon) , \tag{10.17}$$

where $D(\varepsilon)$ is the density of states (DOS) defined by

$$D(\varepsilon) = \sum_{\nu} \delta(\varepsilon - \varepsilon_\nu) , \tag{10.18}$$

while

$$s(\varepsilon) = \log\left[1 + \exp\left(-\beta\left(\varepsilon - \mu\right)\right)\right] . \tag{10.19}$$

We make Chebyshev expansions for $D(\varepsilon)$ and $s(\varepsilon)$ in the forms

$$D(\varepsilon) = \frac{1}{\pi\sqrt{1-\varepsilon^2}} \left[\mu_0 T_0(\varepsilon) + 2 \sum_{m\geq 1} \mu_m T_m(\varepsilon) \right] , \tag{10.20}$$

$$s(\varepsilon) = s_0 T_0(\varepsilon) + 2 \sum_{m\geq 1} s_m T_m(\varepsilon) , \tag{10.21}$$

where Chebyshev moments μ_m and s_m are inversely transformed by

$$\mu_m = \int_{-1}^{1} T_m(\varepsilon) D(\varepsilon) \mathrm{d}\varepsilon = \sum_{\nu} T_m(\varepsilon_\nu), \tag{10.22}$$

$$s_m = \int_{-1}^{1} \frac{\mathrm{d}\varepsilon}{\pi\sqrt{1-\varepsilon^2}} s(\varepsilon) T_m(\varepsilon) . \tag{10.23}$$

From (10.17)–(10.23), we obtain

$$S_{\text{eff}} = -\mu_0 s_0 - 2 \sum_{m\geq 1} \mu_m s_m . \tag{10.24}$$

Namely, instead of using eigenvalues for the Hamiltonian matrix, the effective action can be calculated from the moments $\{\mu_m\}$ and $\{s_m\}$. In practice, we approximate (10.24) by taking finite summations up to $m = M$. We will later discuss about the truncation errors due to finite M in detail.

In (10.22) and (10.23), we have assumed that the absolute values of the eigenvalues are bounded within the range of $|\varepsilon_\nu| \leq 1$. If this is not the case, the Hamiltonian should be rescaled to satisfy the above condition.

Calculation of the DOS Moments. Here we discuss the actual calculation of μ_m. From the property of a matrix trace, we see

$$\mu_m = \sum_{\nu} T_m(\varepsilon_\nu) = \mathrm{Tr}\left\{T_m(H^{\mathrm{e}})\right\} , \tag{10.25}$$

where $T_m(H^e)$ are the matrix polynomials for the Hamiltonian defined by

$$T_0 = 1, \quad T_1 = H^e, \quad T_{m+1}(H) = 2H^e \cdot T_m(H^e) - T_{m-1}(H^e) . \qquad (10.26)$$

Namely, the DOS moments μ_m can simply be obtained by the trace operation without any integrations or matrix diagonalizations. This property gives advantages of the polynomial expansion method as will be discussed later.

Implementation is as follows. Using orthonormal basis vectors $\{\boldsymbol{v}_k^{(0)}\}$, the trace in (10.25) can be calculated by

$$\mu_m = \sum_k \tilde{\mu}_m(k) , \quad \tilde{\mu}_m(k) \equiv (\boldsymbol{v}_k^{(0)}, \boldsymbol{v}_k^{(m)}) , \qquad (10.27)$$

where (,) represents an inner product, and

$$\boldsymbol{v}_k^{(m)} = T_m(H^e)\, \boldsymbol{v}_k^{(0)} . \qquad (10.28)$$

Here, $\boldsymbol{v}_k^{(m)}$ can be obtained recursively by

$$\boldsymbol{v}_k^{(m)} = 2H^e \boldsymbol{v}_k^{(m-1)} - \boldsymbol{v}_k^{(m-2)} , \qquad (10.29)$$

while $\boldsymbol{v}_k^{(1)} = H^e \boldsymbol{v}_k^{(0)}$. Regarding the orthonormal basis vectors, we use unit vectors, i.e., $v_k^{(0)}|_i \equiv \delta_{ik}$, for the simplicity of the calculation.

In practice, relations in (10.15a-b) give more optimized forms for the moment calculation, which are

$$\tilde{\mu}_{2m}(k) = 2(\boldsymbol{v}_k^{(m)}, \boldsymbol{v}_k^{(m)}) - 1 , \qquad (10.30a)$$
$$\tilde{\mu}_{2m+1}(k) = 2(\boldsymbol{v}_k^{(m)}, \boldsymbol{v}_k^{(m+1)}) - (\boldsymbol{v}_k^{(0)}, \boldsymbol{v}_k^{(1)}) . \qquad (10.30b)$$

Namely, when we truncate the Chebyshev series expansion at $m = M$, we only need to calculate $\boldsymbol{v}_k^{(m)}$ up to $m = M/2$. For a system with a particle-hole symmetry, the DOS becomes an even function of energy. Then, odd DOS moments vanishes since the Chebyshev polynomials $T_m(x)$ are odd-functions for odd number of m. Therefore, the calculation for odd moments can be omitted in (10.22).

10.3.2 Advantages of the Polynomial Expansion Method

Accuracies and Stabilities. One of the great advantages to introduce the Chebyshev polynomial expansion is that the moments s_m in (10.23) decay exponentially as m is increased at $m \gg 1$, so that the convergence of S_{eff} is very quick [3]. In other words, a small number of moments M is enough to obtain an accurate result for S_{eff}. It is always possible to make order estimates for the truncation errors by making direct comparisons between moment expansion results and exact results obtained by the diagonalization method, since executions of the Hamiltonian diagonalization for a small number of field configurations $\{x_r\}$ are not beyond the ability of computer resources.

It should also be noted that the recursion formula in (10.13) is known to be numerically stable with the increase of M [4,5]. Therefore, it is guaranteed to improve the accuracy by simply increasing the number of M. In such a sense, the polynomial expansion method is a well-controlled approximation.

In order to calculate the effective action (10.11), the distribution of all the eigenvalues throughout the spectrum bandwidth is necessary. The Chebyshev polynomial expansion gives accurate and stable results for this purpose [4,5]. The Lanczos method or the power method are alternative algorithm to calculate the DOS. However, their accuracies are restricted only to a few absolute largest eigenvalues.

Computational Complexity. The calculation of $\mu_m(k)$ in (10.30a-b) involves matrix-vector products in a recursive way. Computational complexity of a matrix multiplication scales as $O(N)$, if the matrix is sparce and only non-zero elements are treated. In this case, the overall computational complexity for calculating $\{\mu_m\}$ up to $m = M$ is $O(MN)$.

However, in the present case, the computational complexity can further be reduced as follows. We define z as the maximum number of off-diagonal matrix elements of H^{e} for each column. At the beginning, the vector $\boldsymbol{v}_k^{(0)}$ has only one non-zero element. During the recursion process in (10.29), we perform matrix-vector products repeatedly, and each time H^{e} is multiplied, the number of non-zero vector elements increases at most by z. Namely, the number of non-zero elements for $\boldsymbol{v}_k^{(m)}$ at $m \leq M$ scales as $O(M)$. If the matrix-vector product operation is restricted to non-zero vector elements and non-zero matrix elements, the computational complexity of calculating μ_m for $m = 1, \ldots, M$ scales as $O(M^2)$.

If $\tilde{\mu}_m(k)$ is given, the DOS moments μ_m is obtained by the trace operation, or the summation over k in (10.27), which scales as $O(N)$. Then, the calculation of total DOS moments costs $O(M^2N)$, and the overall computational complexity for each local-update MC step scales as $O(M^2N^2)$, since the degrees of freedom for $\{x_r\}$ scales as $O(N)$. Considering that the computational complexity of the diagonalization method scales as $O(N^4)$, the polynomial expansion method is advantageous if $M^2 \ll N^2$ is satisfied.

We see that the value of M can be kept fixed when we increase N [3], which means that the overall computational complexity should be described as $O(N^2)$ for $N \gg 1$. The reason is as follows. For a fixed value of M, relative truncation error of S_{eff} depends on N in general, since the eigenvalues ε_ν and hence μ_m depend on N. One might be concerned that M could increase as a positive power of N under the constraint that the relative truncation error of S_{eff} should be limited to a permissible value. However, this is not the case. We note that $s(\varepsilon)$ and hence s_m are independent of N, and that s_m decays exponentially for large m. We can fix M to a large enough number so that S_{eff} can be calculated with exponentially small truncation errors regardless of N, except for an unusual case where μ_m exponentially diverges at $N \to \infty$.

Let us briefly summarize. Using the Chebyshev polynomial expansion method, the diagonalization process with $O(N^4)$ computational complexity is replaced by the series of matrix multiplications and trace operations, which provide reduced computational complexity of $O(N^2)$ for large enough values of N. Recently, the authors have further extended the algorithm to obtain the DOS moments approximately by truncating the matrix product. The method gives $O(N)$ computational complexity for each local-update MC step, albeit the extra truncation errors for matrix multiplication [7].

Parallel Computation. Matrix operations, in general, have possibilities that they are efficiently performed on parallel computers. In the present case, an implementation for a parallel computation is to distribute the trace operations. In (10.30a-b), the calculation of $\tilde{\mu}_m(k)$ for each k can be performed independently. Therefore, they can be distributed to different nodes of a parallel computer. As an input, each node requires the trial field variables $\{x_r\}$ to construct the Hamiltonian matrix. The output $\tilde{\mu}_m(k)$ should be gathered to obtain μ_m. If the program is coded using the MPI, data communications can simply be realized by `MPI_bcast` and `MPI_reduce` subroutines.

Let us comment on parallelization efficiency for the implementation described above, based on some benchmark results. We consider the case where matrix dimension is $\sim 10^2$ or higher. For shared-memory-type parallel computers, the efficiency is almost perfect. For distributed memory machines, the algorithm suffers the communication overheads, but is still very efficient up to $\sim 10^1$ nodes.

10.4 Application of the Polynomial Expansion Monte Carlo Method

In this section, we demonstrate an example where the polynomial expansion MC method has given an breakthrough for the investigation of an electronic model [6]. The reduction of the CPU time allows us to calculate clusters with larger system sizes. Finite size scaling analysis for a fixed electron density is realized by this method. Not only quantitative increase in size but also qualitative improvements in the comprehension of the model are obtained.

10.4.1 Model

We consider a simplified double-exchange model which has a spin Ising symmetry. Replacing the Heisenberg spins in (10.3) by Ising spins, we have

$$\mathcal{H} = -t \sum_{<ij>,\sigma} c^\dagger_{i\sigma} c_{j\sigma} - J_H \sum_i \sigma^z_i \cdot S_i \,. \qquad (10.31)$$

Here, $c_{i\sigma}$ ($c^\dagger_{i\sigma}$) is an electron operator which annihilates (creates) a σ-spin electron at site i, $\sigma^z_i = n_{i\uparrow} - n_{i\downarrow}$ represents a z-component of the spins of

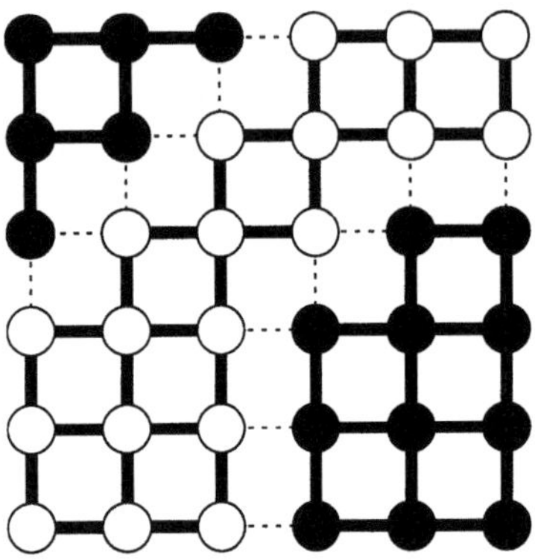

Fig. 10.1. A snapshot of the spin configuration. Filled (open) circles represent sites with $S_i = 1$ ($S_i = -1$). Conduction electrons can hop only between sites connected by bold lines

conduction electrons, and $S_i = \pm 1$ describes the localized Ising spin. Electron hopping energies and Hund's-rule couplings are denoted by t and J_{H}, respectively.

Hereafter, we consider the limit of infinitely-strong Hund's-rule couplings, $J_H \to \infty$, for simplicity. This projects out conduction electrons of which spin states are antiparallel to the localized Ising spins. Then, if $S_i \neq S_j$, an electron at i-th site cannot hop to j-th site since either the initial state or the final state is projected out. In other words, an electron transfer between two sites is allowed only if the Ising spins on these sites are equal. The model can be mapped to a spinless-fermion model with spin-configuration-dependent hopping integrals,

$$\mathcal{H} = -\sum_{<ij>} t(S_i, S_j) \tilde{c}_i^\dagger \tilde{c}_j \,, \quad t(S_i, S_j) = \begin{cases} t, & S_i = S_j \\ 0, & S_i \neq S_j \end{cases} \,. \tag{10.32}$$

Here, spinless-fermion operators $\tilde{c}_i$ and $\tilde{c}_i^\dagger$ denote electronic states parallel to the localized Ising spins.

There is no direct interaction between Ising spins. However, the kinetic energy of conduction electrons depends on the spin configuration. In Fig. 10.1 we illustrate a snapshot of the Ising spin configuration and the conduction electron hoppings for the configuration. We see that the electrons are confined within the ferromagnetic domains.

It is easily understood that the maximum gain of the kinetic energy is realized when the localized Ising spins are perfectly ordered in a ferromagnetic way so that the conduction electrons are maximally delocalized. Namely, ground states of these models are ferromagnetic. This implies that there exist effective ferromagnetic interactions between localized Ising spins. These ferromagnetic interactions share a common origin with those for the ordinary double-exchange model [8,9].

In these models, however, it is difficult to estimate the total kinetic energy for a generic spin configuration. Since kinetic energies depend on sizes and shapes of the domains, effective spin interactions cannot be described in simple analytic forms. They include long range interactions as well as multiple spin interactions.

In this sense, studies of critical phenomena in these models are nontrivial. Since analytic forms of the interactions are not available, MC calculations seem to be one of the most promising ways. However, as described earlier, there exists the difficulty in the diagonalization method that the CPU time diverges to the 4-th power of the system size. The limit for the maximum available lattice size has prevented us from performing finite size analyses. Therefore, application of the polynomial expansion method plays an essential role for investigating the critical phenomena.

10.4.2 Critical Temperature and Critical Exponents

We investigate the ferromagnetism of the two-dimensional simplified double-exchange model in (10.32) at finite temperatures. One of the aim is to determine the universality class and discuss the characters of the ferromagnetic interactions.

MC calculations are performed on clusters with lattice sizes $N = L \times L$ where L are even numbers. We typically run 100 000 MC steps for measurements after 10 000 thermalization steps. At $T \sim T_{\mathrm{c}}$, we find that the calculation of the Chebyshev moments up to $M \sim 20$ is enough to reduce systematic truncation errors so that they become smaller than statistical errors. We take half-bandwidth of the non-interacting system $W \equiv 4t$ as a unit of energy. The only controlling parameter is the electron density n, which is fixed to $n = 0.5$ where the system shows a particle-hole symmetry.

Ferromagnetic order parameter is defined by

$$M_{\mathrm{s}} = \sum_i S_i \,. \tag{10.33}$$

We apply the Binder parameter analysis and the finite size scaling method to calculate the critical temperature T_{c} as well as critical exponents [10].

The Binder parameter is defined by the fourth order cumulant of the order parameter,

$$g = 1 - \frac{\langle M_{\mathrm{s}}^4 \rangle}{3\langle M_{\mathrm{s}}^2 \rangle^2} \,. \tag{10.34}$$

The Binder parameter shows no system size dependence at $T = T_{\mathrm{c}}$. Thus T_{c} is estimated from the crossing point of g as a function of T for various system sizes. Figure 10.2 shows the temperature dependence of g for system

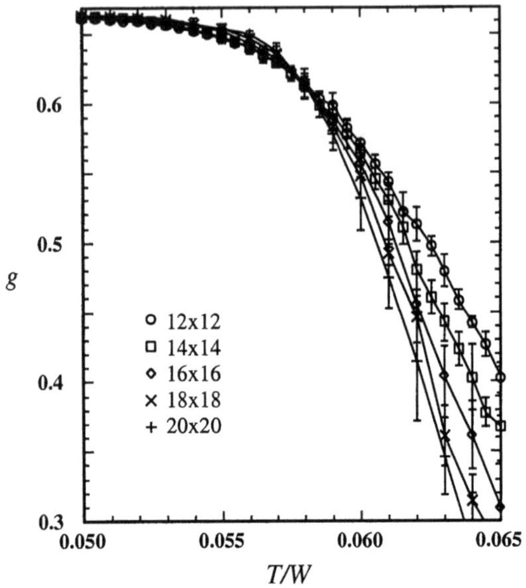

Fig. 10.2. Binder parameter as a function of temperature for various system sizes. The lines are guides to the eye

sizes 12×12 to 20×20. In our model, the Binder parameter seems to cross at $T/W \simeq 0.058$.

Next, we perform a finite-size scaling analysis in the form

$$\chi_s L^{\eta-2} = f(\epsilon L^{1/\nu}), \tag{10.35}$$

where $\chi_s = \langle M_s^2 \rangle / N$ and $\epsilon = (T - T_c)/T_c$, while f is the scaling function. For given exponents η, ν, and T_c, we plot $\chi_s L^{\eta-2}$ as a function of $\epsilon L^{1/\nu}$. If the choice of the exponents and T_c are appropriate, the plotted data converge to a single curve for different values of L and T. The scaling behavior at these values is shown in Fig. 10.3. All the MC data are scaled well to a single universal function within the errors.

As a conclusion from combined analyses of finite size scaling and Binder plots, we obtain $\beta = 0.09 \pm 0.08$, $\nu = 0.9 \pm 0.3$ and $T_c/W = 0.058 \pm 0.001$. The critical exponents of the Ising model with nearest-neighbor interactions in two dimensions are $\beta = 0.125$ and $\nu = 1$. These values agree with our estimates within the error bars. On the contrary, the critical exponents of the mean-field approximation which appear when the interaction is long-ranged, $\beta = \nu = 1/2$, are inconsistent with our data. These results suggest that the universality class of the ferromagnetic transition in our simplified model (10.32) is the same as that of models which have short-range two-spin interactions and the same spin symmetry.

Similar but less accurate results are obtained for the original double-exchange model in three dimensions, i.e., the model in which conduction elec-

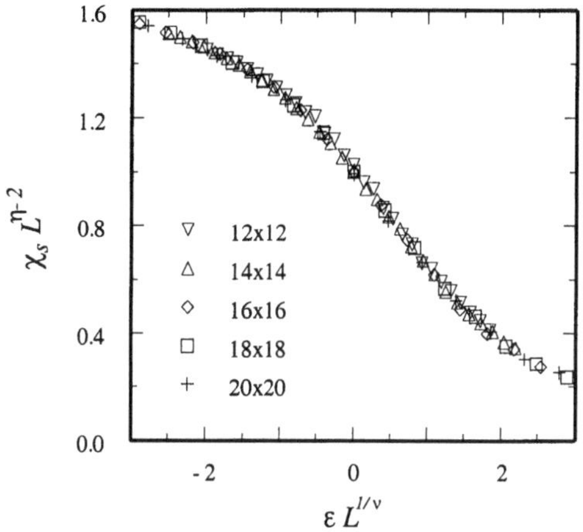

Fig. 10.3. The best-fit result of the finite-size scaling plot for the MC data. The scaling is applied on the data for the systems with $L = 12, 14, 16, 18$ and 20 for $0.05 \leq T/W \leq 0.065$. Error bars are within the sizes of the symbols

trons are coupled with Heisenberg spins [11]. Exponents for the ferromagnetic transition at finite temperature take similar values to those for short-ranged Heisenberg model in three dimensions. From these observations, we speculate that the double-exchange interactions at $T \sim T_c$ are short-ranged, and the multiple-spin interactions are irrelevant.

Up to the best of our knowledge, this is the first example that critical exponents of a finite-temperature magnetic transition in an itinerant electron system are estimated by the finite-size scaling method.

10.5 Summary

To summarize, the algorithm for the MC calculation using the Chebyshev polynomial expansion is as follows.

- At the beginning of the MC calculation, calculate the m-th Chebyshev moment s_m from (10.23) using numerical integration, up to the M-th order.
- Run a standard classical MC for the field variables $\{x_r\}$.
- For each trial field $\{x_r\}$, calculate the Boltzmann weight $P(\{x_r\})$ as follows:
  - Calculate the DOS moment μ_m using (10.30a-b).
  - Calculate S_{eff} by (10.24), and the Boltzmann weight is given by (10.12).

The advantage of the method is that it accelerates the computation, since the computational complexity is reduced and the program runs efficiently on parallel computers. The method has a computational complexity of $O(M^2N^2)$, where N is the system size, whereas the diagonalization method gives $\sim O(N^4)$. The disadvantage is that there exist truncation errors, so that the value of M has to be chosen appropriately for each set of parameters. We should also note that the advantage of the reduced computational complexity is restricted for the cases of large system sizes with $N \gg M$. Therefore, for rather small system sizes $M \sim N$, the simple diagonalization method should be applied.

The method has been used so far to investigate critical phenomena of electron systems with magnetisms. Calculations of electronic properties, such as the DOS as well as various response functions, are also within the scope of the method. It is also interesting to apply the method to other models, such as strongly coupled electron-lattice systems.

Acknowledgment

The authors thank M. Troyer and K. Kubo for fruitful discussions. The computations have been performed mainly on the AOYAMA+ PC cluster systems (`http://www.phys.aoyama.ac.jp/~aoyama+`). Technical assistance by H. Nakata is greatly acknowledged. This work is financially supported by Research Institute for Science and Technology, Aoyama Gakuin University.

References

1. M. Imada, A. Fujimori, Y. Tokura: Rev. Mod. Phys. **70**, 1039 (1998)
2. N. Furukawa, in *Physics of Manganites*, Eds. T. Kaplan and S. Mahanti (Plenum Publishing, New York 1999) and references therein
3. Y. Motome and N. Furukawa: J. Phys. Soc. Jpn. **68**, 3853 (1999)
4. L.W. Wang: Phys. Rev. B **49**, 10154 (1994)
5. R.N. Silver and H. Röder: Int. J. Mod. Phys. C **5**, 735 (1994)
6. Y. Motome and N. Furukawa: J. Phys. Soc. Jpn. **70**, 1487 (2001)
7. N. Furukawa and Y. Motome: submitted to J. Phys. Soc. Jpn.
8. C. Zener: Phys. Rev. **82**, 403 (1951)
9. P.W. Anderson and H. Hasegawa: Phys. Rev. **100**, 675 (1955)
10. D.P. Landau and K. Binder: *A Guide to Monte Carlo Simulations in Statistical Physics* (Cambridge University Press, Cambridge 2000)
11. Y. Motome and N. Furukawa: J. Phys. Soc. Jpn. **69**, 3785 (2000); ibid. **70**, 3186 (2001)

11 From Frustrated Ising Models to Quantum Computing

M. Troyer[1], L.B. Ioffe[2,3], M.V. Feigel'man[3], A. Ioselevich[3], D. Ivanov[1], and G. Blatter[1]

[1] Theoretische Physik, ETH-Hönggerberg, CH-8093 Zürich, Switzerland
[2] Department of Physics and Astronomy, Rutgers University, Piscataway, NJ 08854, USA
[3] Landau Institute for Theoretical Physics, 117940 Moscow, Russia

Abstract. As simulation techniques are maturing, new connection between previously separate fields appear. We present numerical simulations on the quantum dimer model. They show that this model, originally derived as effective model for the low-energy physics of frustrated Ising models, has the right properties to be used in a physical realization of topologically protected quantum bits. A topologically protected quantum bit has the advantage of being passively stable against decoherence and thus does not require error correction schemes.

11.1 Introduction

As the field of simulations in strongly correlated systems, at first motivated by the experimental discovery of heavy fermion systems and high temperature superconductors, is maturing new connection to emerging fields appear. Motivated by the proposal by Kitaev [1] for topologically protected quantum bits, we went on a search for a physically realizable model with the right properties. In this article we present results of our search for a strongly correlated system, exhibiting a two-fold topologically degenerate isolated ground state that is robust with respect to disorder. The model that turned out to have the desired properties is the hardcore dimer model, derived originally as a low-energy effective model for the fully frustrated Ising model. The main results of this work were published recently in a Letter to Nature [2]. In this proceedings we put more emphasis on the simulation techniques and algorithms, an refer to [2] for details of the implementation.

11.2 Quantum Computing

All physical implementations of quantum bits (qubits), carrying the information and computation in a putative quantum computer, have to meet the conflicting requirements of environmental decoupling while remaining manipulable through designed external signals. Proposals based on quantum optics naturally emphasize the aspect of optimal isolation [3–5], while those following the solid state route exploit the variability and scalability of modern nanoscale fabrication techniques [6–10]. Recently, various designs using superconducting structures have been successfully tested for quantum coherent operation [11–13], however, the ultimate goal of reaching coherent evolution over thousands of elementary operations remains a formidable task.

Any quantum computer has to incorporate some fault tolerance as we cannot hope to eliminate all the various sources of decoherence. Amazing progress has been made in the development of quantum error correction schemes [14] involving redundant multi-qubit encoding of the quantum data combined with periodic error detection and recovery steps through appropriate manipulation of the data during execution. Quantum error correction schemes are generic (and hence are applicable to any hardware implementation), but require repeated active interference with the computer during run-time; the delocalization of the data, often in a hierarchical structure, boosts the system size by a factor 10^2 to 10^3. Delocalization of the quantum information is also at the heart of topological quantum computing [1], however, the stabilization against decoherence is entirely deferred to the hardware level (and hence is tied to the specific implementation) and is achieved passively. In searching for a physical implementation of topological qubits one strives for an extended (many body) quantum system where the Hilbert space of quantum states decomposes into mutually orthogonal sectors, each sector remaining isolated under the action of any local perturbation. Choosing the two qubit states from ground states in different sectors protects these states from unwanted mixing through noise; protection from leakage within the sector has to be secured through a gapped excitation spectrum. As no local operator can interfere with these states, global operators must be found (and implemented) allowing for the manipulation of the qubit state.

11.3 Numerical Methods and Tools

11.3.1 Algorithms

In the search for the ideal strongly correlated system, all candidate models need to be probed for the required properties:

1. *A (two-fold) degenerate ground state.* The two degenerate ground states represent the two states of the quantum bit.
2. *A gap separating the degenerate ground state from the rest of the spectrum.*
3. *Robustness of the degeneracy against local disorder.* Small local disorder can never be excluded in the manufacturing process and the device needs to be intrinsically stable against disorder.
4. *Absence of edge states on a system with open boundaries.* We cannot build a periodic system in a physical device.

Exact diagonalization (ED) of small systems, combined with quantum Monte Carlo (QMC) simulations on larger systems (where possible) are the optimal algorithms to answer these questions.

The first question, degeneracy, can often be settled by symmetry arguments, but a direct calculation of the spectrum by ED is needed to rule out additional degeneracies. This can be done on small systems, and the degeneracies will usually stay the same for larger systems.

The presence of a gap is harder to probe for because of finite size effects. While gaps can easily be determined by ED, the gap size has to be extrapolated to infinite system sizes. This poses a problem for ED because of the restriction to small systems. QMC simulations can probe for the absence of long range order, which is a necessary but not a sufficient condition for the existence of a finite gap. Combination of arguments from ED and QMC can give convincing evidence.

Robustness of the degeneracy against disorder can only be probed by ED, since in QMC it is exceedingly difficult to determine the degeneracy of the ground state, or to reliably calculate a small splitting between two topologically degenerate ground states.

The final questions about edge states can again most easily be answered by ED where a direct comparison of the spectra on systems with open and periodic boundaries is possible.

11.3.2 Generic Exact Diagonalization Calculations

In the search for a model with the above properties we do not face the standard problem of determining the properties of a single model, but rather face an inverse problem: what is the model with the envisioned properties? We thus have to be investigate a large number of candidate models and need a simulation code that can easily be adapted to a new model. In the course of the research that lead to this work we have developed a number of generic libraries that will also be useful for future simulations.

Exact diagonalization is just a calculation of the low-lying eigenstates of the Hamiltonian matrix on a finite system. As the Hamiltonian matrix is usually very sparse, iterative eigensolvers are ideal and have been implemented in the Iterative Eigenproblem Template Library [19]. The Hamiltonian itself is usually described in physics in an operator notation, and a library to directly implement this quantum operator notation in C++ was developed. Finally the "lattice" on which the model is defined is a graph in the mathematical sense, often created by periodically translating a unit cell. We have developed an XML file format for such lattices, and a library to create a graph from an XML lattice specification. These libraries will now be presented in more detail.

Iterative Eigenproblem Template Library. Large sparse eigenvalue problems are common to many areas of scientific computing and not just exact diagonalization of quantum systems. Thus it is surprising that despite more than half a century of research on iterative eigensolvers, in-depth reviews and text books appeared only in the last two years [15,16] and comprehensive eigensolver libraries are still missing.

In the related field of linear solvers, the excellent overview of algorithms in the templates book [17] was converted into C++ code by the authors of the Iterative Template Library (ITL) [18]. We have extended the ITL to eigensolvers in the *Iterative Eigenproblem Template Library (IETL)* [19] library. IETL uses an extension of the generic interface of the ITL. Thus it can be used with almost any data type:

- *Any matrix type*: dense, banded, sparse, block-sparse, ... Even distributed matrices on a parallel computer or matrix-free use is possible (calculation of the matrix on-the-fly).
- *Any array type*: dense, sparse, distributed, ...
- *Any scalar type:* single, double, ...precision; real or complex.

This allows the user of the library to employ his favorite matrix and vector types (e.g. BLAS-style arrays, MTL vectors and arrays [21], Blitz++ arrays [22], or even distributed classes) by providing just a few interface functions. In, other libraries (in FORTRAN or C) restrict the choice of vector usually to BLAS-style arrays.

At the moment the Iterative Eigenproblem Template Library (IETL), contains the following algorithms, with more to follow:

- Hermitian and real symmetric matrices:
  - Lanczos algorithm
  - Davidson algorithm
- Nonsymmetric matrices:
  - power method
  - Arnoldi algorithm.

For the simulations presented here the *Lanczos algorithm*, which was implemented following the reference implementation by Cullum and Willoughby [20] was the optimal choice.

Quantum Operator Expression Library. Once an optimal eigensolver, like the Lanczos algorithm is implemented, ED reduces to specifying the Hamilton matrix to be diagonalized. This is usually done in second-quantized operator notation, e.g. as:

$$H = -t \sum_{r=1}^{L} \left(c_r^\dagger c_{r+1} + H.c. \right) + V \sum_{r=1}^{L} n_r n_{r+1} - \mu \sum_{r=1}^{L} n_r \,. \tag{11.1}$$

The implementation of such an operator expression in a simulation program starts by creating a set of basis states, where each is represented by a bit pattern, describing the number of particles on each lattice site. The creation operators $c_r^\dagger$ and annihilation operators c_r then change the bit pattern, while a number operator n_r just measures the number of particles in a state. All of these operations can be implemented in an efficient way using bit operations [23]. This is however a complex and error-prone task and a library to automatically do this is preferred.

Indeed, we have shown that it is possible to do this automatically and efficiently. We have implemented an expression template [24] library for quantum operators on lattices [25]. This library allows to write a matrix-vector multiplication with a second-quantized operator expression:

$$|\phi\rangle = \sum_{r=1}^{L-1} \left[-t\left(c_r^\dagger c_{r+1} + H.c.\right) + V n_r n_{r+1}\right] |\psi\rangle \tag{11.2}$$

simply as

```
Range r(1,L-1);
phi = sum(r,-t*(cdag(r)*c(r+1)+Hc)+V*n(r)*n(r+1))*psi \ts.
```

This allows for the quick-and-easy investigation of a large number of models.

This library is at the moment a prototype[1] but is available to interested readers directly from the author.[2] At the moment the Hamilton operator needs to be fixed at compile-time. This enables compile-time optimization and very efficient codes, comparable to hand-written FORTRAN codes, requires a time-consuming recompilation for each new model. Work on a more flexible, albeit slower, version which reads the model Hamiltonian from an input file is in progress.

Lattice Library and File Format. While the Hamilton operator in our quantum operator library needs to be fixed at compile time, the lattice structure and lattice sizes, which need to be changed more often, can be specified at runtime without loss of efficiency.[3] We can compile our code for a Hamiltonian such as:

$$H = -t \sum_{\langle i,j\rangle} \left(c_i^\dagger c_j + H.c.\right) + V \sum_{\langle i,j\rangle} n_i n_j \,. \tag{11.3}$$

The sum over i and j runs over all neighbors i and j on our finite lattice, which using mathematical terms, is a sum over all edges in a graph.

[1] Only particle number and spin conservation symmetries are implemented. Translational and rotational symmetries can further reduce the size of the Hilbert space but have not been implemented yet.

[2] M. Troyer: troyer@comp-phys.org

[3] The lattice structure enters the simulation only in the outer loops, while the time-consuming inner loops go over the basis states which are independent of the lattice structure.

We have developed an XML format for the specification of lattice graphs [26]. For example, a finite 5×6 square lattice with open boundaries along the first dimension and periodic boundaries in the second dimension, with one vertex (lattice site) per unit cell, and edges to the vertex in all nearest neighbor cells can be specified as:

```
<LATTICEGRAPH>
  <FINITELATTICE>
    <LATTICE dimension="2"/>
    <EXTENT dimension="1" size="5"/>
    <EXTENT dimension="1" size="6"/>
    <BOUNDARY dimension="1" type="open"/>
    <BOUNDARY dimension="2" type="periodic"/>
  </FINITELATTICE>
  <UNITCELL nvertices="1">
    <EDGE><SOURCE vertex="1"><TARGET vertex="1" offset="1 0"></EDGE>
    <EDGE><SOURCE vertex="1"><TARGET vertex="1" offset="0 1"></EDGE>
  </UNITCELL>
</LATTICEGRAPH>
```

We have developed a generic library to read such an XML file and create a graph data structure [27]. Again the user is free to choose his favorite graph data structure and library. By employing the generic interface of the Boost Graph Library [28] the graph is created in the data structure specified by the user. The quantum operator library discussed in the previous section can take this graph as an input and iterates over all the edges in applying the Hamilton operator (11.3).

11.3.3 Performing the Simulations and Parallelization

The combination of above libraries (and additional code in case of the dimer models discussed below, which don't fit the operator representation yet) allows us to quickly simulate a large number of models.

Parallelization can be performed at two levels. First, for a parameter and model search like the one we needed to perform, the required diagonalizations can be farmed out on a cluster, each node working on one specific problem. This approach will work as long as a single diagonalization fits into the memory of a node. As our quantum operator library does not store the matrix but calculates it efficiently on-the-fly during the simulation, memory requirements are reasonably small except for the largest systems.

If an ED simulation no longer fits into the memory of a node, a distributed vector type can be used. The IETL library fully supports distributed vectors and was already used for distributed ED calculations, but this level of parallelism was not needed for the simulations reported here.

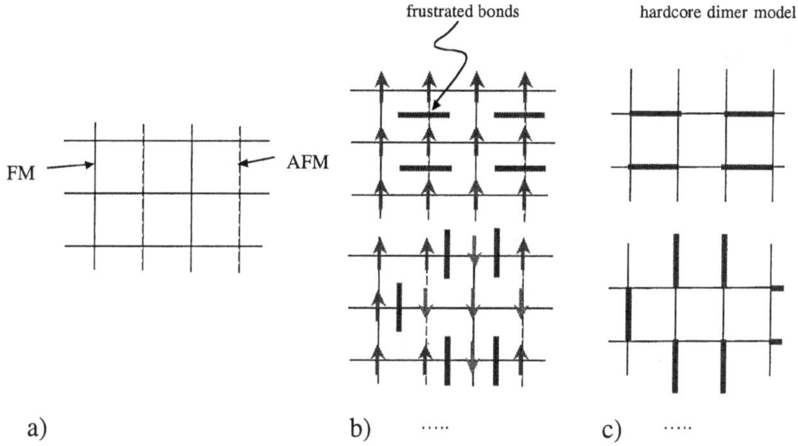

Fig. 11.1. The mapping of the fully frustrated Ising model on a square lattice to a dimer model. (**a**) the square lattice with antiferromagnetic (AFM) bonds indicated as dashed line. (**b**) Two of the macroscopically many ground state configurations. Due to the frustration there is always at least one bond not in its lowest energy configuration on each plaquette, marked by a line across the bond (**c**) the lines across the frustrated bonds make up a full hardcore dimer covering of the dual lattice

11.4 From Frustrated Ising Models to Quantum Dimer Models

After months of simulations we realized that the hardcore quantum dimer system [29–33] is a promising candidate.

This model can be derived as an effective model for the low-energy states in a fully frustrated Ising model

$$H = -\sum_{\langle i,j\rangle} J_{ij}\sigma_i^z\sigma_j^z\,, \tag{11.4}$$

where the product of the signs of the J_{ij} around each plaquette is negative. The simplest case of a fully frustrated square lattice is shown in Fig. 11.1. As each plaquette is frustrated there is at least one bond that is not satisfied. A macroscopically large number of configurations with exactly one non-satisfied bond per plaquette makes up the ground state manifold. By marking these bonds by a dimer on the dual lattice one gets an exact mapping of the ground state manifold to the hardcore dimer coverings of the dual lattice, as shown in Fig. 11.1.

Adding a weak transverse field Γ partially lifts the degeneracy in the ground state manifold by introducing a quantum dynamics:

$$H = -\sum_{\langle i,j\rangle} J_{ij}\sigma_i^z\sigma_j^z + \Gamma\sum_i \sigma_i^x\,. \tag{11.5}$$

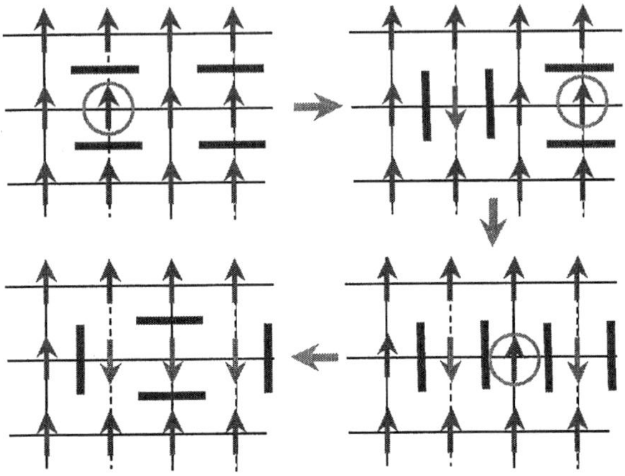

Fig. 11.2. Single-spin flip dynamics of the fully frustrated Ising model in a transverse field

As shown in Fig. 11.2 spins between two parallel dimers can be flipped without leaving the ground state manifold. In terms of the dimers this rotates two parallel dimers on appropriate plaquettes $\square$:

$$H_t = -t \sum_{\square} \left(|\,||\,\rangle\langle\,=\,| + |\,=\,\rangle\langle\,||\,| \right) . \tag{11.6}$$

As a result, the degeneracy of the classical model is lifted, with 'columnar' and 'staggered' (e.g., brickwall) phases (see Fig. 11.3) assuming minimal and maximal energies, respectively. In order to avoid such crystalline ordering we frustrate parallel dimers with the interaction term (which can be derived perturbatively for weak transverse fields $\Gamma \ll |J_{ij}|$ in the Ising model)

$$H_v = v \sum_{\square} \left(|\,||\,\rangle\langle\,||\,| + |\,=\,\rangle\langle\,=\,| \right) . \tag{11.7}$$

While the order characterizing the crystal phases is obvious, the topological order present in the dimer liquid phase is more subtle [31–33]: we consider a cylindrical geometry with periodic boundary conditions along the x-axis, leaving the boundaries along the y-direction free (an additional periodic boundary condition along the y-direction defines a torus). The sum ϕ counting directed dimers cut by the reference line γ parallel to the y-axis defines a topological order parameter (dimers pointing into opposite directions are counted with opposite sign): the action of the Hamiltonian $H = H_t + H_v$ leaves ϕ invariant and the Hilbert space splits into topological sectors $\mathcal{H}_\phi$ characterized by the integers ϕ.

The degeneracy of the ground state is thus lifted. And at the Rokhsar–Kivelson point $t = v$, a zero energy dimer liquid ground state is formed involving all allowed configurations with equal amplitudes [32]. For the square

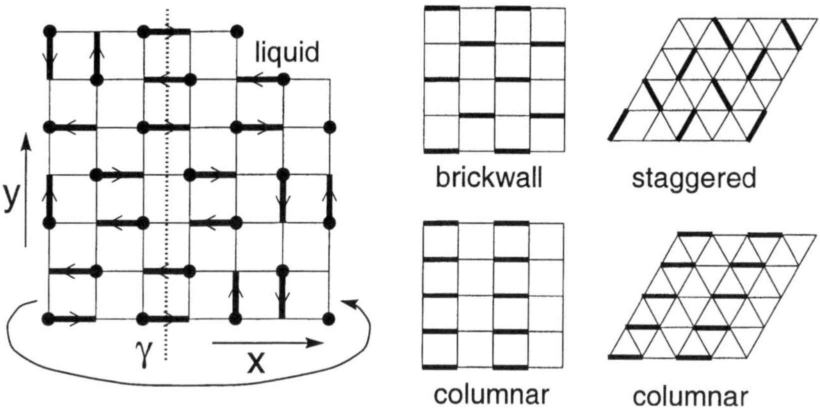

Fig. 11.3. Dimers on a square lattice forming liquid (main panel), brickwall, and columnar configurations. Periodic boundary conditions are applied along the x-direction. The sublattice A is marked by solid bullets; dimers are attached a direction pointing away from bullets. The topological invariant ϕ counts the number of dimers crossing the reference line γ, with dimers pointing to the right (left) counting with positive (negative) signs (here, $\phi = 1$). Right: staggered and columnar configurations in the triangular lattice

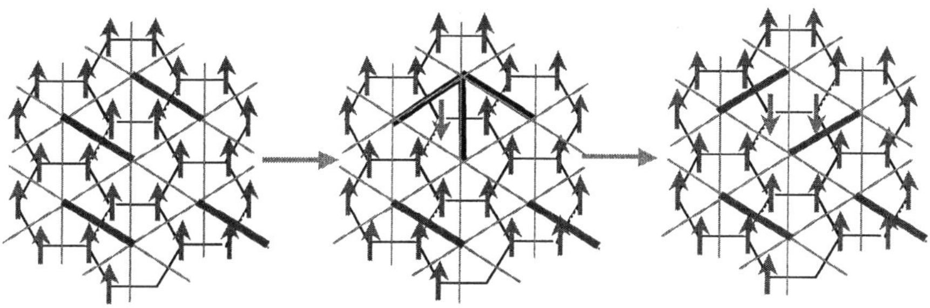

Fig. 11.4. The fully frustrated Ising model on a hexagonal lattice. As in the case of the square lattice the ground states can be mapped onto the hardcore dimer coverings of the dual (triangular lattice). In the quantum dynamics single spin flips leave the ground state manifold, but two-spin flips can stay in the ground state manifold, again giving rise to rotaton of dimers on a plaquette

lattice this liquid phase is however restricted to the point $t = v$ and tuning v away from t the liquid loses the competition with various crystalline phases [34]. As we are interested in systems developing a stable liquid over an extended parameter range we have to continue our search.

Recent quantum Monte Carlo simulations of the dimer model on a triangular lattice by Moessner and Sondhi [29] provide evidence that the corresponding model on a triangular lattice (derived from the fully frustrated Ising model on the hexagonal lattice, see Fig. 11.4) has a gapped liquid ground state in an extended parameter regime.

Let us first investigate the topological properties of this model. As the triangular lattice is no longer bipartite, we cannot introduce directed dimers pointing from one sublattice to the other. We thus do not have an infinite number of topological sectors. Instead the dimer dynamics only changes the number of dimers cutting the reference line γ by 0 or ± 2, conserving the parity. Thus in going from the square lattice to the triangular lattice the infinite number of topological sectors collapse to two sectors, with odd or even numbers of dimers crossing the reference line.

In the thermodynamic limit $L_y \to \infty$ the gap between the ground states of the two topological sectors vanishes exponentially fast and we end up with two *topologically degenerate ground states*, which we confirmed in an exact diagonalization calculation.

The next property to check for is the existence of a gap above this two-fold degenerate ground state. Indeed, this expectation is supported by results obtained from numerical studies: recent Monte Carlo simulations by Moessner and Sondhi [29] on large systems with $L_{x,y} = 36$ exhibit short-range dimer-dimer correlations within a parameter region $2/3 < v/t < 1$, indicative of a *liquid state*. Furthermore, the weak temperature dependence of this data within the temperature interval $0.25t > T > 0.03t$ provides evidence for a gapped spectrum (increasing v beyond t, a first-order transition takes the liquid into a staggered phase; decreasing v below $2t/3$ establishes intermediate crystalline order with a $\sqrt{12} * \sqrt{12}$ unit cell, before the system settles into the columnar phase at small v).

While these results are clear evidence of a liquid state, they do not exclude the possibility of having additional low energy liquid-like states below the gap top the crystalline phases. We have thus numerically diagonalized the t-v dimer Hamiltonian $H = H_t + H_v$ on a triangular lattice in order to estimate the gap protecting the liquid and verify the absence of low lying edge states; in addition we have investigated the robustness of the degenerate ground states under local perturbations (requiring exact diagonalization limited to smaller systems). We have chosen both, cylindrical and toroidal geometries, with system sizes going up to $L_{x,y} = 6$; although subject to finite size effects we are confident that our results describe reasonably well the liquid phase where the short-ranged dimer correlations (of order of one lattice constant [29]) reduce the influence of the boundaries.

Our exact diagonalization study, see Fig. 11.5, confirms the *presence of a gap* of order $\Delta \approx 0.1t$, taking up large values on approaching the Rokhsar-Kivelson point $t = v$ (the largest gap of order $0.2t$ appears in the $6 * 3$ torus at $v/t \approx 0.8$; the distance between levels just above the gap is $\sim 0.01\,t$ and falls rapidly for higher energies). Comparing the spectra for tori (no open boundaries) with those of cylinders (where the two open boundaries could accommodate edge states, would they exist) we note that the gap persists, from which we infer the absence of low lying edge-states in the dimer liquid (this should be contrasted with the situation in the quantum Hall system).

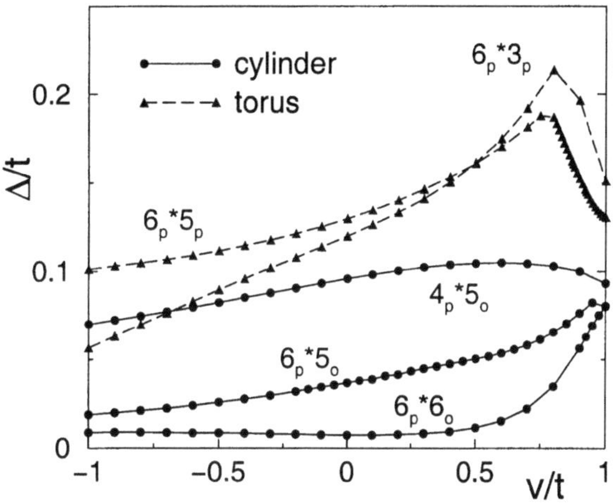

Fig. 11.5. Excitation gap Δ protecting degenerate ground states for cylinders $4_p * 5_o$, $6_p * 5_o$, $6_p * 6_o$, and for tori $6_p * 3_p$ and $6_p * 5_p$. With increasing v the system evolves from fluctuating columnar to liquid to static staggered phases, with a sharp crossover at the Rokhsar-Kivelson point $v = t$. Approaching the latter from small v we find the largest gap of order $0.2t$ in the $6_p * 3_p$ torus with a generic four-fold degenerate ground state

In order to test the susceptibility of the degenerate ground states to local perturbations we assign all links random chemical potentials μ_d homogeneously distributed over the interval $[-d/2, d/2]$, $d < t$. As shown in Fig. 11.6 the ground state splitting Δ_d collapses dramatically as we go from the crystal ($v > t$) to the liquid phase ($v < t$) and slowly recovers as v is decreased further. We attribute the sharp drop at $v \approx t$ to the first-order solid-liquid transition; the small disorder splitting at $v < t$ testifies for the efficient *protection* of the dimer liquid ground states *from local perturbations.* Decreasing v further below t we observe various signatures which we attribute to the appearance of intermediate crystalline phases [29] before reaching the fluctuating columnar phase at large negative values of v.

11.5 From Dimer Models to Quantum Computing

Summarizing, the combination of Monte Carlo simulations [29] and diagonalization indicate that on approaching the point $v = t$ from below, the triangular dimer model develops an *isolated degenerate dimer liquid ground state free of edge states.* Within the parameter region $\sim 0.8 < v/t < 1$ we find a gap of order $0.1t$ and ground states which are *robust under disorder* to within 1% of the disorder potential; we expect a further increase of this robustness with system size.

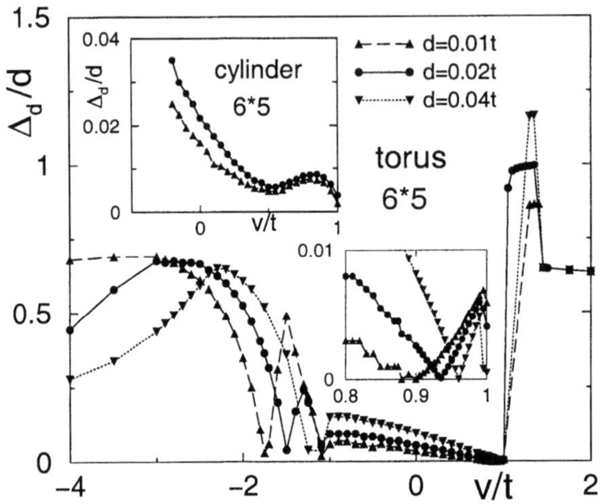

Fig. 11.6. Splitting Δ_d of the ground state energies under the action of a disorder potential of strength d for a $6*5$ torus and a $6*5$ cylinder (inset; periodicity is along x with $L_x = 6$). The disorder only weakly perturbs the dimer liquid ground states at $v < t$ (note the expanded view near $v \sim t$ for the $6 * 5$ torus). The crossovers between various phases become clearly visible in the susceptibility to a disorder potential

Translating these findings to potential implementations of this two-state system, we find the latter to be quite tolerant with respect to (static) local variations in parameters appearing in the fabrication process. Regarding decoherence, the amplitude mixing of the two ground states is inhibited through their topological nature and the diffusion in their relative phase is suppressed, again a consequence of the robustness with respect to (dynamical) local disorder. Excitations within each topological sector are inhibited by the gap Δ. The latter sets the requirements on the operation temperature: in order to avoid excitations, the system has to be cooled to a temperature below $T^* \sim \Delta / \ln N_{\text{cell}}$, with N_{cell} the number of cells in the array. Here, the gap energy Δ is an order of magnitude below the energy scale t set by the device.

The physical implementation of such a dimer system can be realized with the help of quantum Josephson junction arrays. Such arrays allow for a proper 'material design' as the energy scales governing the charge and phase degrees of freedom can be tuned through junctions with appropriate tunnel barriers and capacitances. Shaped into the form of a ring, the Josephson junction array emulating a dimer system allows for the construction of topologically protected qubits. We refer to the Nature publication [2] for details of the implementation.

11.6 Outlook

As our knowledge of exotic quantum states grows, new connection to other fields develop. In this work we could span a bridge from frustrated Ising models, an old field of research to quantum dimer models and topological order in quantum system, a relatively new field. The goal of this study, the development of a topologically protected quantum bit, brings in a third very active field of research.

The new design for an unconditionally and passively stable quantum bit could play an important role in the development of a quantum computer, which could help solve many currently unsolvable computational problems.

For the short term however a quantum computer will not be realized. There is however progress also on classical computers. Since the algorithms for the simulation of quantum and classical systems have reached a reasonable level of maturity we can now start the development of standardized simulation codes. Such standardized codes, which have been popular in other fields of computational sciences, like quantum chemistry or computational fluid dynamics, could also profit researchers in the field of strongly correlated systems. The large number of interesting models and algorithms makes the development of these codes a big challenge.

Acknowledgements

We thank V. Geshkenbein for discussions and acknowledge financial support through the SCOPES program (swiss federal department of foreign affairs and SNF), the Swiss National Science Foundation, the NWO-Russia collaboration program, the RFBR grant 01-02-17759, the program 'Quantum Macrophysics' of the Russian Academy of Science, and the Russian Ministry of Science. Computations have been carried out on the Beowulf cluster Asgard at ETHZ.

References

1. A.Yu. Kitaev: *Fault-tolerant quantum computation by anyons.* Report quant-ph/9707021
2. L. Ioffe, M.V. Feigel'man, A. Ioselevich, D. Ivanov, M. Troyer and G. Blatter: Nature **415**, 507 (2002)
3. J.I. Cirac and P. Zoller: Phys. Rev. Lett. **74**, 4091 (1995)
4. C. Monroe, D. Meekhof, B. King, W. Itano and D. Wineland: Phys. Rev. Lett. **75**, 4714 (1995)
5. Q. Turchette, C. Hood, W. Lange, H. Mabushi, and H.J. Kimble: Phys. Rev. Lett. **75**, 4710 (1995)
6. D. Loss and D.P. DiVincenzo: Phys. Rev. A **57**, 120 (1998)
7. A. Shnirman, G. Schön, and Z. Hermon: Phys. Rev. Lett. **79**, 2371 (1997)
8. D.V. Averin: Solid State Commun. **105**, 659 (1998)

9. J.E. Mooij, T.P. Orlando, L.S. Levitov, L. Tian, C.H. van der Wal, C.H. and S. Lloyd: Science **235**, 1036 (1999)
10. L. Ioffe, V.B. Geshkenbein, M.V. Feigel'man, A.L. Fauchère and G. Blatter: Nature **398**, 678 (1999)
11. Y. Nakamura, Yu.A. Pashkin and J.S.Tsai: Nature **398**, 786-788 (1999)
12. J.R. Friedman, V. Patel, W. Chen, S.K. Tolpygo and J.E. Lukens: Nature **406**, 43 (2000)
13. C.H. van der Wal, A.C.J. ter Haar, F.K. Wilhelm, R.N. Schouten, C.J.P.M. Harmans, T.P. Orlando, S. Lloyd, and J.E. Mooij: Science **290**, 773 (2000)
14. J. Preskill: *Fault-tolerant quantum computation.*
15. Z. Bai, J. Demmel and J. Dongarra (Eds.), *Templates for the Solution of Algebraic Eigenvalue Problems: A Practical Guide* (SIAM, 2000)
16. G.W. Stewart: *Matrix Algorithms* (SIAM, 2001)
17. R. Barret, M. Berry, T.F. Chan, J. Demmel, J. Donato, J. Dongarra, V. Eijkhout, R. Pozo, C. Romine, and H. van der Vorst: *Templates for the Solution of Linear Systems: Building Blocks for Iterative Methods* (SIAM, 1993)
18. J.G. Siek, A. Lumsdaine and L.-Q. Lee, in *Proceedings of the SIAM Workshop on Object Oriented Methods for Inter-operable Scientific and Engineering Computing (OO'98)* (SIAM, 1998); the library is availavle on the web at: `http://www.osl.iu.edu/research/itl/`
19. `http://www.comp-phys.org/software/ietl/`
20. J.K. Cullum and R.A. Willoughby: J. Comput. Phys. **44**, 329–358 (1981); J.K. Cullum and R.A. Willoughby: *Lanczos algorithms for Large Symmetric Eigenvalue Computations.* Vols. 1 and 2, Theory. (Birkhauser, Boston 1985)
21. J.G. Siek and A. Lumsdaine, in *Modern Software Tools for Scientific Computing* (Birkhäuser, Boston 1999); the library is availavle on the web at: `http://www.osl.iu.edu/research/mtl/`
22. T. Veldhuizen, in *Lect. Notes in Computer Science* **1505** 223 (1998); the library is available on the web at `http://www.oonumerics.org/blitz/`
23. M. Troyer, Ph.D. thesis (ETH Zürich, 1994)
24. T. Veldhuizen, C++ Report **7**, 26 (1995); A.D. Robinson: Computers in Physics **10**, 458 (1996)
25. M. Troyer: Lect. Notes in Computer Science **1732**, 164 (1999)
26. See the documentation at `http://xml.comp-phys.org`
27. `http://www.comp-phys.org/software/PALM++/lattice/`
28. L.Q. Lee, A. Lumsdaine and J.G. Siek: *The Boost Graph Library* (Addison Wesley, 2001)
29. R. Moessner, and S.L. Sondhi: Phys. Rev. Lett. **86**, 1881 (2001)
30. R. Moessner, and S.L. Sondhi: Phys. Rev. B **63**, 224401 (2001)
31. S.A. Kivelson, D.S. Rokhsar, and J.P. Sethna: Phys. Rev. B **35**, 8865 (1987)
32. D.S. Rokhsar, and S.A. Kivelson: Phys. Rev. Lett. **61**, 2376 (1988)
33. X.G. Wen: Phys. Rev. B **44**, 2664 (1991)
34. E. Fradkin: *Field Theories of Condensed Matter Systems.* (Addison Wesley, Redwood City, CA 1991)

12 Parallel Quantum Monte Carlo Simulation of $S = 3$ Antiferromagnetic Heisenberg Chain

S. Todo*

Theoretische Physik, Eidgenössische Technische Hochschule,
CH-8093 Zürich, Switzerland
Institute for Solid State Physics, University of Tokyo, Kashiwa 277-8581, Japan

12.1 Introduction

Since Haldane made a striking conjecture that the integer-spin ($S = 1$, 2, 3, ...) antiferromagnetic Heisenberg chain has a finite excitation gap above its unique ground state [1], estimation of its precise value has been one of the most challenging problems in the computational condensed-matter physics. For the spin-1 system the finiteness of the first excitation gap as well as the correlation length has been established numerically until early 90s [2]. Simulations of higher-spin systems, however, are much harder since the magnitudes of the inversed gap and the correlation length would increase exponentially as S increases. Indeed the asymptotic forms for the large-S limit [1] suggests that in the spin-3 case they are of order 10^2 and 10^3, respectively.

The main difficulty for measuring such a small gap is that the system looks gapless (critical) as long as the temperature is not low enough compared to the gap [3]. Similarly the system size used in the simulate should be much larger than the correlation length to detect the finite correlation length. Thus we need to simulate quite large systems ($\sim 10^4$) at extremely low temperatures ($\sim 10^{-3}$) in order to observe ground-state quantities in the thermodynamic limit.

In the present paper we report the results of our recent large-scale quantum Monte Carlo (QMC) simulation for $S = 1$, 2, and 3 antiferromagnetic Heisenberg chains [4]. For the successful simulation of the spin-3 system, the following ingredients are essential: (i) efficient QMC dynamics by using the continuous-time loop algorithm, (ii) improved higher-order gap estimator, and (iii) parallelization on massively parallel computers. In the following sections we will discuss each point. Especially in Sect. 12.4 we will give a detailed explanation on the parallelization technique of the continuous-time loop algorithm.

---

* Present address: Department of Applied Physics, University of Tokyo, Tokyo 113-8656, Japan

12.2 Continuous-Time Loop Algorithm

The *loop algorithm*, invented in early 1990s, is a variant of the cluster algorithm for QMC simulations [5]. In the cluster algorithm, instead of local updates of spin configuration, spins belonging to the same cluster are flipped simultaneously. In the case of quantum Heisenberg antiferromagnets, each cluster forms a one-dimensional object, which is referred to as a *loop*.

One Monte Carlo step of the loop algorithm consists of two procedures: the labeling process and the loop-identification one. In the former process graphs are assigned stochastically to each plaquette of the $(d+1)$-dimensional lattice, so that flipping spins connected by an edge of the graph does not break the local conservation law of S^z in the imaginary-time direction. The essential point is that the assignment of graphs is done only by using the local spin configuration on each plaquette.

Next the $(d+1)$-dimensional lattice is decomposed into a set of loops based on the graphs assigned to the plaquettes. Then spins on each loop are flipped as a whole according to a certain probability. The labeling and flipping probabilities should be chosen so that the detailed-balance condition is fulfilled for the entire process. The loop algorithm is ergodic and in addition it reduces the auto-correlation time greatly, sometimes by orders of magnitude.

Furthermore one can formulate the loop algorithm directly in the imaginary-time continuum. This method is called the *continuous-time loop algorithm* [6]. In this algorithm the Trotter limit is taken prior to the simulation and thus the results is completely free from the Trotter discretization error.

For systems with $S \geq 1$, each spin is represent as a sum of $2S$ spin-$\frac{1}{2}$ spins (subspins) [4,7]. Simultaneously each bond is transformed into $4S^2$ bonds of the same strength as the original one. For the mapped system the spin-1/2 continuous-time loop algorithm can be used in a straight forward way. Note that in order to recover the dimension of the Hilbert space of the original system one needs to introduce a set of projection operators, each of which acts on $2S$ subspins and projects out unphysical states with $\boldsymbol{S}^2 < S(S+1)$. In the path-integral representation, the projection operators are transformed into special boundary conditions in the imaginary-time direction [4,7].

12.3 Improved Gap Estimator

In the path-integral representation, the first excitation gap Δ at momentum π is given by the inverse of the (dominant) correlation length ξ_τ of staggered magnetization in the imaginary-time direction. However, extracting the correlation length from the correlation function, which is equivalent to the inverse Laplace transform problem, is extremely difficult, since in general it's quite numerically unstable.

On the other hand, the correlation length obtained by the standard second-moment method [8], which is frequently used for studying second-order phase transitions, is suffered from systematic errors, since this method gives us not an *exponential* correlation length, but an *integrated* one. For example, for spin-1 system the relative systematic error is about 0.2%, which is much larger than the typical statistical error.

In order to remove this systematic error, we proposed the higher-order moment estimators [4], by which one can reduce the error in a systematic way. We find that the convergence of the result in terms of the order of estimator is quite rapid and typically the systematic error in the fourth-order moment estimator becomes smaller than the statistical one.

12.4 Parallelization

For the $S = 3$ chain we need $L = 5792$ and $T = 0.0010359$ to reach the thermodynamic limit as we will see in Sect. 12.5. The QMC simulation of the largest system requires more than 6 GB memory and it does not fit into the memory available on a typical processor (or node). In addition, one Monte Carlo step would take about one hour on a single CPU. Therefore, we need to parallelize each QMC step. In general, however, the parallelization of the loop algorithm is highly non-trivial, since it constructs global objects (loops) and tries to flip them as a whole.

In order to parallelize the loop algorithm, first we need to determine how to distribute the information of a spin (world-line) configuration to several nodes. In the continuous-time loop algorithm it might be natural to divide the $(d+1)$-dimensional space into N_p slices of thickness β/N_p in the imaginary-time direction (Fig. 12.1), where N_p is the number of nodes, since the system is continuous in this direction and therefore the thickness of each slice can be the same regardless of N_p. Another great advantage of this scheme is that the algorithm does not depend on the dimensionality of the system we simulate.

The labeling process can be trivially parallelized, since it requires only local information, i.e. the relative direction of two neighboring spins at the same imaginary time. No communication between nodes is required at this stage. On the other hand, however, the loop-identification process should be preformed in two steps: (i) each processor identifies loops in its imaginary-time window. Although in this step almost all the loops are closed, there remain $N \times N_\mathrm{p}$ unclosed loops, where N is the number of subspins, since they cross the imaginary-time boundaries between the nodes. This stage also requires no inter-node communication. Then (ii) these unclosed loops are merged into large closed loops. This step can be done quite efficiently by using the binary-tree algorithm as shown in Fig. 12.2. Finally the information on global loops are distributed back to all the nodes to determine the next spin configuration.

The number of operations required for the global loop-merging process does not depend on the temperature. It is proportional to the number of subspins N and $\log N_\mathrm{p}$. On the other hand, that for the other processes is

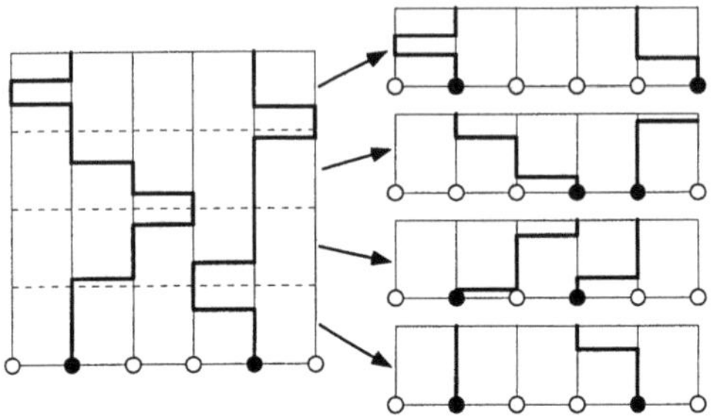

Fig. 12.1. Decomposition of a world-line configuration into 4 nodes. Horizontal (vertical) axis denotes the real-space (imaginary-time) direction. Each node stores the spin directions at imaginary time $\beta p/N_{\rm p}$, where p is the node index, and the positions of jumps of world-lines in its own time window

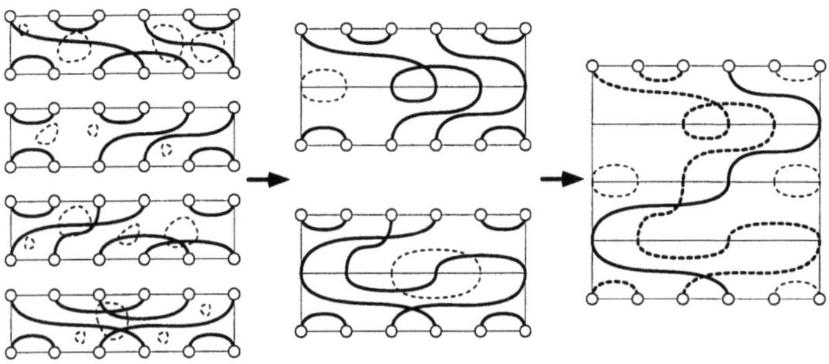

Fig. 12.2. Schematic picture of the loop-merging process for $N_{\rm p} = 4$. Thin dashed lines denote locally-closed loops, while two thick (solid and dashed) lines in the rightmost figure denote global loops constructed at the final stage

proportional to $N\beta$ in total, where β is the inverse temperature $(1/T)$. Since the latter one can be performed independently on $N_{\rm p}$ nodes, overall overhead for parallelization is proportional to $N_{\rm p} \log N_{\rm p}/\beta$ for large $\beta/N_{\rm p}$. The one Monte Carlo step for the spin-3 chain with $L = 5792$ and $T = 0.0010359$, which is mapped to a system of 34752 subspins and 208512 bonds prior to the simulation, takes about 16 s on 256 nodes of Hitachi SR-2201.

12.5 Results

In Fig. 12.3, we present the temperature dependence of the correlation length in the imaginary-time direction ξ_τ as well as that in the spatial direction ξ_x. Note that the ratio of the inverse temperature $\beta = 1/T$ to the system size L is fixed to a constant $(1/2S)$. At high temperatures, $\xi_\tau(L,\beta)$ and $\xi_x(L,\beta)$ grow linearly as β increases. Once both β and L exceed enough the correlation lengths at zero temperature in the thermodynamic limit, however, $\xi_\tau(L,\beta)$ and $\xi_x(L,\beta)$ no longer exhibit any temperature dependence besides statistical

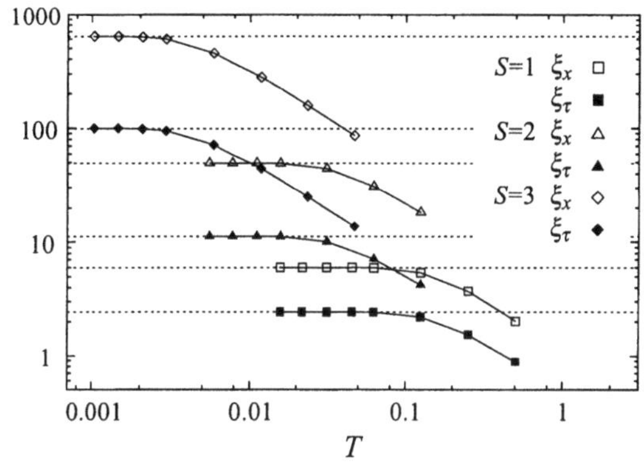

Fig. 12.3. Temperature dependence of the real-space and the imaginary-time correlation lengths of the $S = 1$, 2, and 3 AF Heisenberg chains [4]. Statistical error of each data point is smaller then the line width. The system size L is taken as $2S/T$

fluctuations. Empirically, we find that $L/\xi_x(L,\beta) > 6$ with $\beta/\xi_\tau(L,\beta) > 6$ is a reasonable condition to guarantee the convergence to the thermodynamic and zero-temperature limits.

Thus we conclude that $\Delta = \xi_\tau^{-1} = 0.41048(6)$, 0.08917(4), and 0.01002(3) and $\xi_x = 6.0153(3)$, 49.49(1), and 637(1) for $S = 1$, 2, and 3, respectively. We used the fourth-order estimator for the final results. Note that these figures are taken from the QMC simulation for the largest system at the lowest temperature without any extrapolation procedure. For other quantities see the table given in [4].

For the spin-1 case, the present results agree quite well with the previous estimates by the density renormalization group (DMRG) method and by the exact diagonalization [2]. For larger S, however, these methods become less effective. Indeed the spatial correlation length of the spin-2 system, $\xi_x = 49.49(1)$, is much larger than the system size available in the exact diagonalization method. For $S = 3$, it might be practically impossible to estimate the gap even by the DMRG method. Our result for the spin-3 system is complete new to the best of our knowledge.

Acknowledgement

Most of the numerical calculations in the present work were performed on CP-PACS at the University of Tsukuba and Hitachi SR-2201 at the Supercomputer Center, University of Tokyo. The program used in the present simulation was based on the library "Looper version 2" developed by S.T. and K. Kato. This work is supported by the "Research for the Future Program" (JSPS-RFTF97P01103) of Japan Society for the Promotion of Science and by the Swiss National Science Foundation.

References

1. F.D.M. Haldane: Phys. Lett. **93A**, 464 (1983); Phys. Rev. Lett. **50**, 1153 (1983)
2. See, e.g. S.R. White and D.A. Huse: Phys. Rev. B **48**, 3844 (1993); O. Golinelli, Th. Jolicœur, and R. Lacaze: Phys. Rev. B **50**, 3037 (1994)
3. See, e.g. S. Todo, M. Matsumoto, C. Yasuda, and H. Takayama: Phys. Rev. B **64**, 224412 (2002)
4. S. Todo and K. Kato: Phys. Rev. Lett. **87**, 047203 (2001)
5. H.G. Evertz, G. Lana, and M. Marcu: Phys. Rev. Lett. **70**, 875 (1993); U.-J. Wiese and H.-P. Ying: Z. Phys. B **93**, 147 (1994)
6. B.B. Beard and U.-J. Wiese: Phys. Rev. Lett. **77**, 5130 (1996)
7. N. Kawashima and J.E. Gubernatis: Phys. Rev. Lett. **73**, 1295 (1994); J. Stat. Phys. **80**, 169 (1995); K. Harada, M. Troyer, and N. Kawashima: J. Phys. Soc. Jpn. **67**, 1130 (1998)
8. F. Cooper, B. Freedman, and D. Preston: Nucl. Phys. B **210** [FS6], 210 (1982)

13 Melting of Stripes in a Two-Dimensional Hard-Core Boson Hubbard Model

G. Schmid[1], S. Todo[1], M. Troyer[1], and G. Batrouni[2]

[1] Theoretische Physik, Eidgenössische Technische Hochschule Zürich, CH-8093 Zürich, Switzerland
[2] Institut Non-Linéaire de Nice, Université de Nice-Sophia Antipolis, 1361 route des Lucioles, 06560 Valbonne, France

Abstract. We investigate the melting of the stripe phase in the square lattice hard-core boson Hubbard model with nearest and next nearest neighbor repulsion using quantum Monte Carlo simulations. We find that the stripe melting is realized in a first order transition at low temperatures. In addition we conjecture that a nematic phase does not exist in this model.

A nearly universal feature of strongly correlated systems is a phase transition between a correlation-induced insulating phase with localized charge carriers, to an itinerant phase. High temperature superconductors [1], manganites [2] and the controversial two dimensional (2D) "metal-insulator-transition" [3] are just a few examples of this phenomenon. The 2D hard-core boson Hubbard model provides the simplest example of such a transition from a correlation-induced charge density wave insulator to a superfluid (SF) [4]. It is a prototypical model for preformed Cooper pairs [5], spin flops in anisotropic quantum magnets [6,7], SF Helium films [8] and supersolids [9,10].

In simulations on this model, which does not suffer from the negative sign problem of fermionic simulations, we can investigate some of the pertinent questions about such phase transitions: what is the order of the quantum phase transitions in the ground state and the finite temperature phase transitions? Are there special points with dynamically enhanced symmetry [11]? Can there be coexistence of two types of order (such as a supersolid – coexisting solid and SF order)? Answers to these questions provide insight also for the other problems alluded to above.

The Hamiltonian of the hard-core boson Hubbard model we study is

$$H = -t \sum_{\langle i,j \rangle} (a_i^\dagger a_j + a_j^\dagger a_i) + V_1 \sum_{\langle i,j \rangle} n_i n_j + V_2 \sum_{\langle\langle i,j \rangle\rangle} n_i n_j - \mu \sum_i n_i \, , \quad (13.1)$$

where $a_i^\dagger (a_i)$ is the creation (annihilation) operator for hard-core bosons, $n_i = a_i^\dagger a_i$ the number operator, V_1 and V_2 are the nearest and next nearest neighbor Coulomb repulsions, respectively and μ the chemical potential. This model is equivalent to an anisotropic spin-1/2 XXZ model with $J_{z,1} = V_1$, $J_{z,2} = V_2$ and $|J_{xy}| = 2t$ in a magnetic field $h = 2(V_1 + V_2) - \mu$. A rich

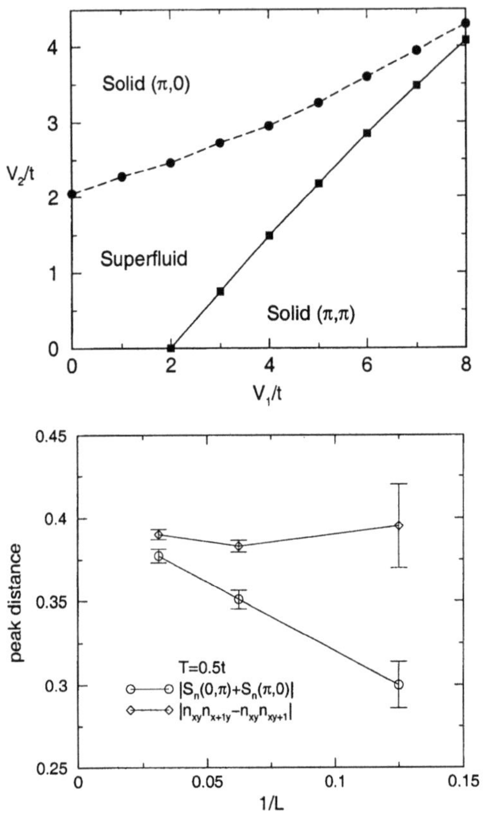

Fig. 13.1. Left: Ground-state phase diagram of the hard-core boson Hubbard model (13.1) at half filling [12]. Right: Finite size scaling for the distance of the peaks in histograms for $|O_N|$ and $|O_S|$ at the transition line using $T = 0.5t$ and $V_1 = 0$. It clearly confirms a first order transition up to $T = 0.5t$

finite-temperature phase diagram of this model at $V_2 = 0$ which shows e.g. a direct first order transition from SF to solid, a tricritical point and unusual reentrant behavior is discussed in [4]. In this report we take into account also next nearest neighbor repulsion but restrict ourselves to zero field (and zero magnetization $m_z = 0$) which corresponds to the half filled bosonic model (density $\rho = 1/2$) at $\mu = 2(V_1+V_2)$. Figure 13.1 shows the ground-state phase diagram [12] of this model. For dominating repulsion V_1 the system shows staggered checkerboard charge order, while it is a striped solid for dominating repulsion V_2. These solid phases are separated from each other by a SF.

All of these phases extend to finite temperatures. In the limit of infinite repulsion V_2 the lattice decouples into two sub-lattices. On either of them the hard-core boson Hubbard model is equivalent to an antiferromagnetic Ising model at $t = 0$ and the insulating behavior extends up to a finite-temperature phase transition of the Ising universality class.

Throughout this report, we focus our attention on the melting of the striped solid at $V_1 = 0$ and $V_2 \gtrsim 2t$. The striped solid has broken rotational and translational symmetry and there are scenarios possible for restoring these symmetries upon heating which have more than one transition line. One possibility are two second order transitions at slightly different temperatures, the lower one breaks the translational and the upper one the rotational symmetry. Between this two transition lines we would then have a nematic phase which has only rotational symmetry broken similar to a liquid crystal.

In order to decide whether a nematic phase plays a role in the melting of the stripes and to determine the finite-temperature phase diagram we use

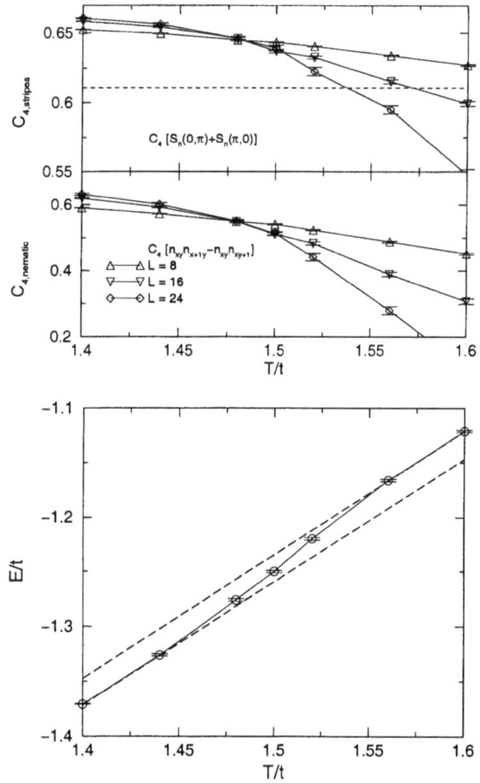

Fig. 13.2. Left: fourth order cumulants for O_s and O_N as function of the temperature ($V_1 = 0$, $V_2 = 3t$). The horizontal line shows the critical value C_4^c for a second order transition in the Ising universality class. The crossing point of the fourth order cumulants for O_S is not consistent with this value. Right: Energy per lattice site as function of the temperature (lattice size $L = 24$, $V_1 = 0$, $V_2 = 3t$). It indicates a jump across the phase transition

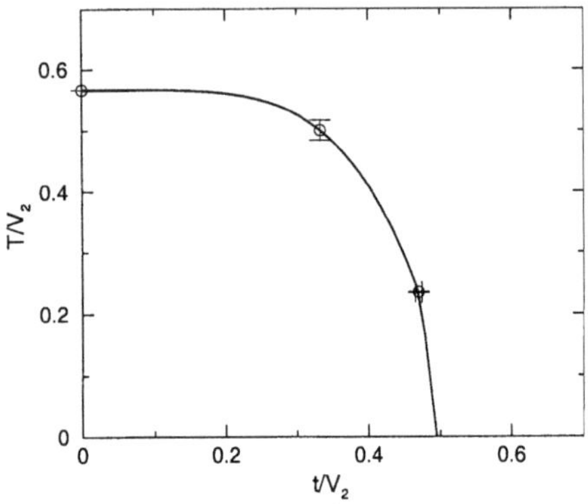

Fig. 13.3. First order melting line for the stripe phase in the hard-core boson Hubbard model (13.1) at $V_1 = 0$. The line is only a guide to the eye

quantum Monte Carlo simulations, employing a variant [13,14] of the worm algorithm [15] in the stochastic series expansion (SSE) representation [13]. The calculations do not suffer from any systematic error apart from finite size effects. In spite of the fact that we perform simulations at half filling (what corresponds to $h = 0$ in the spin language) we did not use cluster updates due to problems with freezing in this frustrated model.

To check for broken symmetries, we use the order parameters $O_S = S_n(\pi, 0) + S_n(0, \pi)$ ($S_n(\boldsymbol{k})$ is the charge structure factor at wave vector $\mathbf{k}$) for the stripe order and $O_N = n_{xy}n_{x+1y} - n_{xy}n_{xy+1}$ ($n_{xy} = a^{\dagger}_{xy}a_{xy}$ is the number operator at position (x, y) in the 2D lattice) for the nematic order.

At low temperature $T = 0.5t$ in the vicinity of the first order quantum phase transition between striped solid and SF, the histograms for both the order parameter $|O_N|$ and $|O_S|$ show two clearly separated peaks what indicates a first order transition. Since a two-peak structure can also be seen at a second order transition in a finite system, a careful finite size scaling analysis for the peak distance is needed. Figure 13.1 shows this finite size scaling for the peak distance. It confirms a melting in a first order transition up to $T = 0.5t$.

For larger next nearest neighbor repulsion $V_2 = 3t$ and correspondingly higher temperature, the classification of the stripe melting is much more difficult. A clear two peak structure could not be seen in the histograms for lattice sizes up to 32×32. To check if there are two second order lines we use a finite size scaling for the fourth order cumulant ratios $C_{4,i} = 1 - \langle O_i^4 \rangle / 3 \langle O_i^2 \rangle^2$ with $i \in \{N, S\}$. Figure 13.2 shows that both, the fourth order cumulants for O_N and O_S seem to have size independent values and thus a phase transition at the same temperature. In a scenario with two separated second order transition lines, the transition to the striped solid should be in the Ising universality class and we would expect for $C_{4,S}$ a size independent universal value of $C_4^c = 0.6106900(1)$ [16], what seems inconsistent with our results.

Both facts suggest that the stripes melt either in a single non-universal second order transition or in a single first order transition. To decide between these two possibilities we calculate the energy per lattice site as function of the temperature. As shown in Fig. 13.2, the energy could have a jump across the transition. Preliminary results for the finite size scaling suggest a finite jump also in the infinite lattice. This indicates that the melting of the stripes is realized by a single first order transition as shown in Fig. 13.3. We conjecture that a nematic phase does not exist in this model.

Acknowledgements

The simulations were performed on the Asgard Beowulf cluster at ETH Zürich, employing the Alea library for Monte Carlo simulations [17]. G.S., S.T. and M.T. acknowledge support of the Swiss National Science Foundation.

References

1. J.G. Bednorz and K.A. Müller: Z. Phys. B **64**, 189 (1986)
2. E. Dagotto et al.: Phys. Rep. **344**, 1 (2001)
3. E. Abrahams et al.: Rev. Mod. Phys. **73**, 251 (2001)
4. G. Schmid et al.: Phys. Rev. Lett. **88**, 167208 (2002)
5. T. Siller et al.: Phys. Rev. B **63**, 195106 (2001)
6. A. van Otterlo et al.: Phys. Rev. B **52**, 16176 (1995)
7. M. Kohno and M. Takahashi: Phys. Rev. B **56**, 3212 (1997)
8. G.T. Zimanyi et al.: Phys. Rev. B **50**, 6515 (1994)
9. G.G. Batrouni et al.: Phys. Rev. Lett. **74**, 2527 (1995); R.T. Scalettar et al.: Phys. Rev. B **51**, 8467 (1995)
10. G.G. Batrouni and R.T. Scalettar: Phys. Rev. Lett. **84**, 1599 (2000)
11. M.E. Fisher and D.R. Nelson: Phys. Rev. Lett. **32**, 1350 (1974)
12. F. Hebert et al.: Phys. Rev. B **65**, 014513 (2002)
13. A.W. Sandvik: Phys. Rev. B **59**, 14157 (1999)
14. A. Dorneich and M. Troyer: Phys. Rev. E **64**, 066701 (2001)
15. N.V. Prokof'ev et al.: Sov. Phys. - JETP **87**, 310 (1998)
16. G. Kamieniarz and H. Blöte: J. Phys. A: Math Gen. **26**, 201 (1993)
17. M. Troyer et al.: Lec. Notes Comp. Sci. **1505**, 191 (1998)

14 A Microscopic Approach to the Mott–Hubbard Gap

C.-H. Pao[1] and N.E. Bickers[2]

[1] Department of Physics, Nat'l Chung Cheng University, Chiayi, Taiwan, 621, ROC
[2] Department of Physics, University of Southern California, Los Angeles, CA 90089, USA

Abstract. We have studied the doped Mott insulator phase of the two-dimension Hubbard model from the intermediate to large interaction regimes. By introducing a rotationally invariant Stratonovich–Hubbard field we decouple the static spin components of the interaction. The static charge interaction is then treated by Hartree–Fock approximation in the presence of a random spin field. We sample random spin configurations by Monte Carlo simulation with the Metropolis updating algorithm. The Mott–Hubbard gap and static susceptibilities are studied at and away from half-filling at low temperature.

We present a new microscopic approach to perturbation theory for systems with local moments. In the past decade, weak-coupling perturbation theories have been used extensively to study the physics of the Hubbard model. In the large U limit at half-filling, the static spin response is unstable within Hartree–Fock and the system exhibits a gap of order U in the one-body density of states. At finite temperature, this long-range antiferromagnetic phase persists to a temperature of order U in the large-U limit. Away from half-filling, the zero-temperature behavior of the Hartree–Fock approximation has also been investigated for incommensurate states containing vertical and diagonal domain walls [1]. However, the competition between different states remains unresolved.

In order to model spin fluctuations in a disordered system, one of the authors (NEB) proposed the fluctuation exchange, or FLEX, approximation [2–4]. The self-consistent growth of the one-body self-energy suppresses the magnetic ordering transition to zero temperature. However, this approach does not show a sharp gap at half-filling in the density of states and leaves a finite weight at zero energy for all temperatures [3–5]. This failure at half-filling brings into question results obtained for large U at small doping, and the FLEX approximation fails to describe the pseudogap regime in the 2D Hubbard model. This inadequacy has motivated various approaches to investigate this pseudogap problem [6–8]. In addition, a technique based on infinite-order (or "parquet") vertex renormalization, which extends the FLEX approximation, has also been applied to this problem [9]. However, the full numerical solutions of the minimal parquet equations for the 2D Hubbard model in the large-U limit show that (i) no gap appears at half filling; and (ii) the spectral density at low energies is actually larger than within the FLEX calculation. This failure strongly suggests that conventional perturbation theory is not suitable for the description of a system with local moments.

The failure of conventional perturbation theory is due to a qualitative change in the effective free-energy which onsets at a temperature of order U. This change comes from the appearance of local moments in the system and depends on both the interaction strength U and the chemical potential μ. One way to approach this problem is to introduce a spatially varying static field with the symmetry of the classical order parameter. In this paper we present such field via a Stratonovich–Hubbard (S–H) transformation [10] of the interaction. In particular this spin-rotation-invariant S–H field decouples the static component of the Hubbard interaction. When the fermion contribution to the action is integrated out, the effective action for the static field reduces to a classical Heisenberg form at half-filling in the large-U limit. In the following paragraph we describe this new approach briefly [11].

The partition function for the two dimensional Hubbard model is

$$Z = \mathrm{Tr} e^{-\beta \hat{H}} , \tag{14.1}$$

where

$$\hat{H} = -t \sum_{\langle ij\rangle\sigma} (c^{\dagger}_{i\sigma} c_{j\sigma} + HC) \; + \; U \sum_{i} n_{i\uparrow} n_{i\downarrow} \; - \; \mu \sum_{i\sigma} n_{i\sigma} \, . \tag{14.2}$$

The sum on $\langle ij\rangle$ is restricted to nearest neighbor pairs, with hopping integral t. U is an intra-orbital (i.e., on-site) Coulomb repulsion, and μ is the chemical potential. It is convenient to introduce an anticommuting c-number notation to approach the problem. We may write the fermion action as

$$S = S_0 + S_{\mathrm{int}} , \tag{14.3}$$

$$S_0 = \beta \sum_{ijn\sigma} \bar{c}_n^{i\sigma} \left[-i\omega_n \mathbf{1} + (\boldsymbol{H}_0 - \mu \mathbf{1}) \right]_{ij} c_n^{j\sigma} , \tag{14.4}$$

$$S_{\mathrm{int}} = -\beta U \sum_{imnn'} \bar{c}_{n+m}^{i\uparrow} c_n^{i\downarrow} \bar{c}_{n'}^{i\downarrow} c_{n'+m}^{i\uparrow} = S_{\mathrm{int}}^{m=0} + S_{\mathrm{int}}^{m\neq 0} , \tag{14.5}$$

with $\omega_n = (2n+1)\pi T$ and

$$\mathbf{1}_{ij} = \delta_{ij}$$

$$(\boldsymbol{H}_0)_{ij} = \begin{cases} -t \, , \; i \text{ and } j \text{ near-neighbors} \\ 0 \; , \; \text{ else} \end{cases} . \tag{14.6}$$

The static interaction ($S_{\mathrm{int}}^{m=0}$) may be decomposed in the charge (repulsive) and spin (attractive) channels as

$$S_{\mathrm{int}}^{m=0} = \frac{1}{4} \beta U \sum_i (n_0^i)^2 - \frac{1}{4} \beta U \sum_i \boldsymbol{m}_0^i \cdot \boldsymbol{m}_0^i , \tag{14.7}$$

where $n_0^i = \sum_{n\sigma} \bar{c}_n^{i\sigma} c_n^{i\sigma}$ and $\boldsymbol{m}_0^i = \sum_{n\alpha\alpha'} \bar{c}_n^{i\alpha} (\boldsymbol{\sigma})_{\alpha\alpha'} c_n^{i\sigma'}$. We may introduce an independent static S-H field at each site i to decouple the spin interaction:

$$\exp[-S_{\text{spin}}^{m=0}] = \int \prod_i \left(\frac{\beta}{\pi}\right)^{3/2} d\boldsymbol{H}^i$$
$$\times \exp\left[-\beta \sum_i \boldsymbol{H}^i \cdot \boldsymbol{H}^i + \beta\sqrt{U} \sum_i \boldsymbol{H}^i \cdot \boldsymbol{m}_0^i\right]. \tag{14.8}$$

Then a Hartree–Fock approximation is applied to the static charge interaction for fixed S–H field. The static action may be written as:

$$\bar{S}(H) = \beta \sum_i |H^i|^2 - \frac{1}{4}\beta U \sum_i \langle n^i \rangle_H^2 + \bar{S}_f(H, \bar{c}, c), \tag{14.9}$$

$$\bar{S}_f(H, \bar{c}, c) = \beta \sum_{ijn\alpha\alpha'} \bar{c}_n^{i\alpha} \Bigg\{ -i\omega_n \delta_{ij} \delta_{\alpha\alpha'} + \Bigg[(\boldsymbol{H}_0)_{ij} \delta_{\alpha\alpha'} - \sqrt{U} \boldsymbol{H}^i \cdot \boldsymbol{\sigma}_{\alpha\alpha'} \delta_{ij}$$
$$+ \left(\frac{1}{2} U \langle n^i \rangle_H - \mu\right) \delta_{ij} \delta_{\alpha\alpha'} \Bigg] \Bigg\} c_n^{j\alpha'}. \tag{14.10}$$

The fermion action (14.10), which involves only the one-body terms, may now be integrated out exactly:

$$\int D\bar{c} Dc \exp[-\bar{S}_f] = \prod_n \left[1 + e^{-\beta\epsilon_n(H)}\right], \tag{14.11}$$

where the ϵ_n are eigenvalues of the one-body Hamiltonian corresponding to the fermion action in (14.10). The effective action can then be obtained after the fermion integration

$$S_{\text{eff}}(H) = \bar{S}_0(H) - \sum_n \log\left[1 + e^{-\beta\epsilon_n(H)}\right]. \tag{14.12}$$

We then evaluate electron correlation function using S_{eff} as the action in a classical field Monte Carlo. This avoids the sign problem inherent in a full quantum Monte Carlo. However, the current theory still has several complications compared to a classical Heisenberg spin model [12]: (i) The current theory has three degrees of freedom instead of two. (ii) Local field updates induce global changes through the fermion integration in (14.11), which involves the diagonalization of the effective Hamiltonian matrix with row dimension $2N_{\text{sites}}$. (iii) A Hartree–Fock convergence step is required after each spin field update. (iv) A Kawasaki dynamics (or spin swapping) is required for an efficient equilibration away from half-fillings.

The data reported below are calculated for 8×8 periodic lattices, Coulomb repulsion $U/t = 8$, and temperature $T/t = 0.125$. There are 100 Monte Carlo warmup sweeps and 5000 sampling sweeps for each runs.

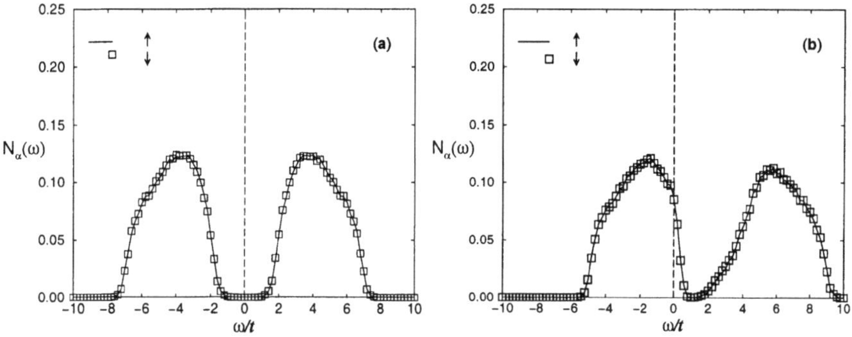

Fig. 14.1. Total density of states for spin α, $N_\alpha(\omega)$. The solid line is for spin up, and the squares are for spin down. The model parameters are $U/t = 8$ and $T/t = 0.125$. (**a**) $\langle n \rangle = 1.0$. (**b**) $\langle n \rangle = 0.94$

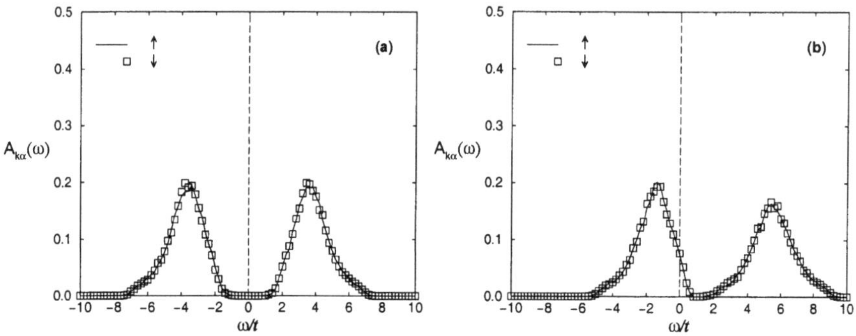

Fig. 14.2. Momentum-resolved spectral density $A_{k\alpha}(\omega)$. The model parameters are the same as in Fig. 14.1. The momentum point is $k = (0, \pi)$. (**a**) $\langle n \rangle = 1.0$. (**b**) $\langle n \rangle = 0.93$

The total density of states for spin α, N_α is plotted in Fig. 14.1. At half filling (Fig. 14.1a), a Mott–Hubbard gap (or pseudogap) appears clearly in the density of states. Away from half-filling, the remnant of this gap remains in Fig. 14.1b. This behavior is consistent with previous Lanczos [13,14] and quantum Monte Carlo studies [15] of the Hubbard model [16]. These results also indicate that the systems are not magnetically ordered, because the spin-up and spin-down densities are identical within statistical sampling errors.

The k-resolved spectral densities are shown in Figs. 14.2 and 14.3 for two different wave vectors $\boldsymbol{k} = (0, \pi)$ (near the Fermi surface) and $\boldsymbol{k} = (\pi, \pi)$ (at the top of the band). In Fig. 14.2a, the spectral weight is split symmetrically at half-filling and the gap is order of U between the upper and lower Hubbard bands. When the system is doped (Fig. 14.2b), the Fermi level moves into the lower Hubbard band. The gap exists for a wide range of fillings for this model parameter but the distribution of the spectral weight changes. For the (π, π) point, the weight is concentrated in the upper Hubbard band as expected.

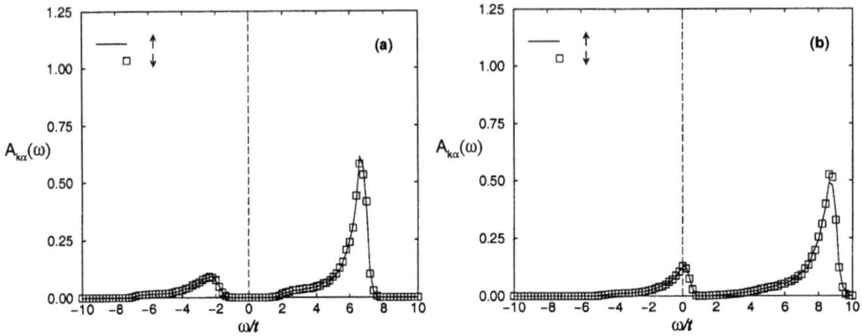

Fig. 14.3. Momentum-resolved spectral density $A_{k\alpha}(\omega)$. The model parameters are the same as in Fig. 14.2. The momentum point is $k = (\pi, \pi)$. (**a**) $\langle n \rangle = 1.0$. (**b**) $\langle n \rangle = 0.93$

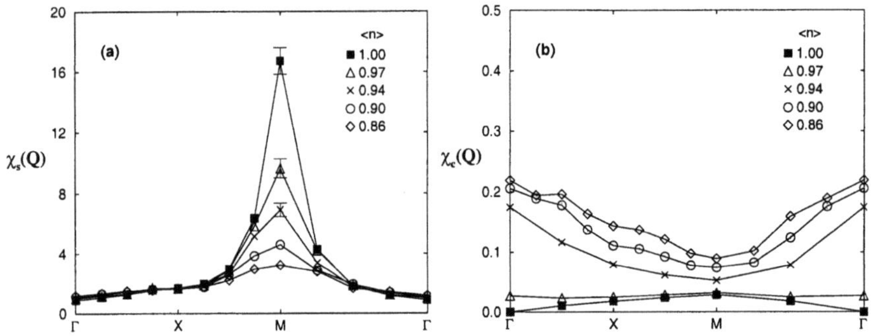

Fig. 14.4. Static (**a**) spin susceptibility $\chi_s(Q)$ and (**b**) charge susceptibility $\chi_c(Q)$ for $U/t = 8.0$ and $T/t = 0.125$. The susceptibilities are plotted along the triangular Brillouin zone contour from $\Gamma \to X \to M \to \Gamma$, i.e., from $(0, 0) \to (\pi, 0) \to (\pi, \pi) \to (0, 0)$. Unless shown, statistical error bars are smaller than the plotting symbols

The static susceptibilities are plotted in Fig. 14.4. In Fig. 14.4a, the static spin susceptibility is large and peaks at the antiferromagnetic wave vector (π, π). This (π, π) peak decreases as the system is doped. Upon doping, we expect the peak to move away from (π, π), at least in large systems. We were not able to observe this behavior for the temperature range and system size discussed here. A larger system would presumably show effects of domain wall formation and incommensurability.

In conclusion we have developed a new method to study the 2D Hubbard model based on Monte Carlo simulation of a classical local spin field. The Mott–Hubbard gap is described at and away half-filling in the present static field analysis.

Acknowledgements

This work was supported by the National Science Council (Taiwan, ROC) under Grant No. NSC 90-2112-M-194-002 (CHP), and by the National Science Foundation (USA) under Grant No. DMR98-11381 (NEB). Computing support from the Nat'l Center for High-Performance Computing (NCHC, Taiwan, ROC) and the computer center of Nat'l Chung Cheng University are acknowledged.

References

1. J. Zaanen and O. Gunnarsson: Phys. Rev. B **40**, 7391 (1989); D. Poilblanc and T.M. Rice: Phys. Rev. B **39**, 9749 (1989); H.J. Schulz: J. Physique **50**, 2833 (1989)
2. N.E. Bickers and D.J. Scalapino: Ann. Phys. (N.Y.) **193**, 206 (1989)
3. N.E. Bickers, D.J. Scalapino and S.R. White: Phys. Rev. Lett. **62**, 961 (1989)
4. N.E. Bickers and S.R. White: Phys. Rev. B **43**, 8044 (1991)
5. S. Moukouri et al.: Phys. Rev. B **61**, 7887 (2000)
6. A. Kampf and J.R. Schrieffer: Phys. Rev. B **42**, 7967 (1990)
7. Y.M. Vilk and A.-M.S. Tremblay: J. Phys. **17**, 1309 (1997)
8. A.V. Chubukov and J. Schmalian: Phys. Rev. B **57**, R11085 (1998)
9. N.E. Bickers submitted to Phys. Rev. B (2000)
10. R.L. Stratonovich: Dokl. Akad. Nauk SSSR **115**, 1097 (1957) [Soviet Phys. Doklady **2**, 416 (1958)]
11. N.E. Bickers and D.J. Scalapino: cond-mat/0010480 (2000)
12. For a more detail description, see [11]
13. E. Dagotto et al.: Phys. Rev. Lett. **67**, 1918 (1991)
14. E. Dagotto, F. Ortolani, and D.J. Scalapino: Phys. Rev. B **46**, 3183 (1992)
15. N. Bulut, D.J. Scalapino, and S.R. White: Phys. Rev. B **47**, 6157 (1993); ibid. **50**, 9623 (1994)
16. For a survey of Lanczos and quantum Monte Carlo results for the Hubbard model, see E. Dagotto: Rev. Mod. Phys. **66**, 763 (1994)

15 Two Quantum Cluster Approximations

T.A. Maier[1], O. Gonzalez[1,2], M. Jarrell[1], and T. Schulthess[2]

[1] University of Cincinnati
Cincinnati OH 45221, USA
[2] Computational Material Sciences Group
Oak Ridge National Laboratory
Oak Ridge, TN 37831, USA

Abstract. We provide microscopic diagrammatic derivations of the Molecular Coherent Potential Approximation (MCA) and Dynamical Cluster Approximation (DCA) and show that both are Φ-derivable. The MCA (DCA) maps the lattice onto a self-consistently embedded cluster with open (periodic) boundary conditions, and therefore violates (preserves) the translational symmetry of the original lattice. As a consequence of the boundary conditions, the MCA (DCA) converges slowly (quickly) with corrections $\mathcal{O}(1/L_c)$ ($\mathcal{O}(1/L_c^2)$), where L_c is the linear size of the cluster. However, local quantities, when measured in the center of the MCA cluster, converge more quickly than the DCA result. These results are demonstrated numerically for the one-dimensional symmetric Falicov–Kimball model.

15.1 Introduction

One of the most active areas in condensed matter physics is the search for new methods to treat chemically disordered and correlated systems. In these systems, especially in three dimensions or higher, approximations which neglect long ranged correlations are generally thought to provide a reasonable first approximation for many properties.

Perhaps the most successful of these methods are the Coherent Potential Approximation (CPA) [1] and the Dynamical Mean Field Approximation (DMFA) [2–5], for disordered and correlated systems, respectively. Although these approximations have different origins, they are formally related. Both are single site theories where non-local effects of disorder and correlations are treated in a mean field approximation. Diagrammatically, both the DMFA [3] and the CPA [6] may be defined as theories which completely neglect momentum conservation at all internal vertices. When this principle is applied, the diagrammatic expansion for the irreducible quantities in each approximation collapses onto that of a self-consistently embedded impurity problem.

Many researchers have actively searched for a technique to restore nonlocal corrections to these approaches. Here, we discuss just two approaches which are fully causal and self-consistent: the Molecular Coherent Potential Approximation (MCA) [7,8] and the Dynamical Cluster Approximation (DCA) [9–11,6]. Recently the Cellular Dynamical Mean Field Approach [12] was proposed for ordered correlated systems, while the Molecular Coherent

Potential Approximation has traditionally been applied to disordered systems. Since both methods share a common microscopic definition we use the term MCA to refer to both techniques in the following.

While the MCA is traditionally defined in the real space of the lattice, the DCA is traditionally defined in its reciprocal space. In the MCA, the system lattice is split into a series of identical molecules. Interactions between the molecules are treated in a mean-field approximation, while interactions within the molecule are explicitly accounted for. In the DCA, the reciprocal space of the lattice is split into cells, and momentum conservation is neglected for momentum transfers within each cell while it is (partially) conserved for transfers between the cells. These approximations share many features in common: they both map the lattice problem onto that of a self-consistently embedded cluster problem. Both recover the single site approximation (CPA or DMFA) when the cluster size reduces to one and become exact as the cluster size diverges. Both are fully causal [7,10], and provided that the clusters are chosen correctly [6], they maintain the point group symmetry of the original lattice problem. Here, we provide a microscopic diagrammatic derivation of both the MCA and the DCA, and explore their convergence with increasing cluster size.

15.2 Formalism

We will employ a diagrammatic formalism to derive the MCA and DCA, assuming that a collection of electrons on a lattice, with Green function $G(\boldsymbol{k})$ interact through an interaction $V(\boldsymbol{k})$.

15.2.1 From the Lattice to the Cluster

Since our object is to define cluster methods, we divide the original lattice of N sites into N/N_c clusters (molecules), each composed of $N_c = L_c^D$ sites, where D is the dimensionality. L_c need not be an integer, for example in a two-dimensional square lattice, a diamond cluster with $N_c = 8$ will have $L_c = 2\sqrt{2}$. However, care must be taken so that the clusters preserve the point group symmetry of the original lattice. We use the coordinate $\tilde{\boldsymbol{x}}$ to label the origin of the clusters and $\boldsymbol{X}$ to label the N_c sites within a cluster, so that the site indices of the original lattice $\boldsymbol{x} = \boldsymbol{X} + \tilde{\boldsymbol{x}}$. The points $\tilde{\boldsymbol{x}}$ form a lattice with a reciprocal space labeled by $\tilde{\boldsymbol{k}}$. The reciprocal space corresponding to the sites $\boldsymbol{X}$ within a cluster shall be labeled $\boldsymbol{K}$, with $K_\alpha = n_\alpha \cdot 2\pi/L_c$ and integer n_α. Then $\boldsymbol{k} = \boldsymbol{K} + \tilde{\boldsymbol{k}}$. Note that $e^{i\boldsymbol{K}\cdot\tilde{\boldsymbol{x}}} = 1$ since a component of $\tilde{\boldsymbol{x}}$ must take the form $m_\alpha L_c$ with integer m_α.

The formal mapping between the full lattice problem and the cluster is accomplished by relaxing the condition of momentum conservation at the

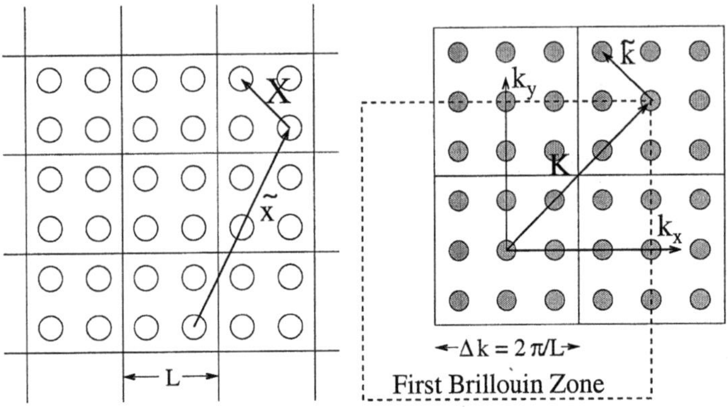

Fig. 15.1. Definitions of the parameters in real (left) and reciprocal (right) space

internal vertices of the compact diagrams. Momentum conservation at each vertex is described by the Laue function

$$\Delta = \sum_{\boldsymbol{x}} e^{i\boldsymbol{x}\cdot(\boldsymbol{k}_1+\boldsymbol{k}_2+\cdots,-\boldsymbol{k}_1'-\boldsymbol{k}_2'-\cdots)} = N\delta_{\boldsymbol{k}_1+\boldsymbol{k}_2+\cdots,\boldsymbol{k}_1'+\boldsymbol{k}_2'+\cdots}\,, \tag{15.1}$$

where $\boldsymbol{k}_1$, $\boldsymbol{k}_2$ ($\boldsymbol{k}_1'$, $\boldsymbol{k}_2'$) are the momenta entering (leaving) the vertex. Müller–Hartmann [3] showed that the Dynamical Mean Field (DMF) theory may be derived by completely ignoring momentum conservation at each internal vertex by setting $\Delta = 1$. Then, one may freely sum over all of the internal momentum labels, and the graphs for the generating functional Φ and its irreducible derivatives, contain only local propagators and interactions.

The DCA and MCA techniques may also be defined by their respective Laue functions. In the MCA, we approximate the Laue function by

$$\Delta_{\mathrm{MC}} = \sum_{\boldsymbol{X}} e^{i\boldsymbol{X}\cdot(\boldsymbol{K}_1+\tilde{\boldsymbol{k}}_1+\boldsymbol{K}_2+\tilde{\boldsymbol{k}}_2+\cdots-\boldsymbol{K}_1'-\tilde{\boldsymbol{k}}_1'-\boldsymbol{K}_2'-\tilde{\boldsymbol{k}}_2'-\cdots)}\,. \tag{15.2}$$

Thus, the MCA omits the phase factors $e^{i\tilde{\boldsymbol{k}}\cdot\tilde{\boldsymbol{x}}}$ resulting from the position of the cluster in the original lattice but retains the (far less important) phase factors $e^{i\tilde{\boldsymbol{k}}\cdot\boldsymbol{X}}$ associated with the position within a cluster. In the DCA we also omit the phase factors $e^{i\tilde{\boldsymbol{k}}\cdot\boldsymbol{X}}$, so that

$$\Delta_{\mathrm{DC}} = N_c\delta_{\boldsymbol{K}_1+\boldsymbol{K}_2+\cdots,\boldsymbol{K}_1'+\boldsymbol{K}_2'+\cdots}\,. \tag{15.3}$$

Both the MCA and DCA Laue functions recover the exact result when $N_c \to \infty$ and the DMFA result, $\Delta = 1$, when $N_c = 1$.

If we apply the MCA Laue function (15.2) to diagrams in Φ, assuming that V is a two-particle interaction then each Green function leg is replaced by the MCA coarse-grained Green function (we have dropped the frequency dependence for notational convenience)

$$\bar{G}(\boldsymbol{X}_1, \boldsymbol{X}_2; \tilde{\boldsymbol{x}} = 0)$$

$$= \frac{1}{N^2} \sum_{\substack{K_1, K_2 \\ \tilde{k}_1, \tilde{k}_2}} e^{i(\boldsymbol{K}_1+\tilde{\boldsymbol{k}}_1)\cdot \boldsymbol{X}_1} G(\boldsymbol{K}_1, \boldsymbol{K}_2; \tilde{\boldsymbol{k}}_1, \tilde{\boldsymbol{k}}_2) e^{-i(\boldsymbol{K}_2+\tilde{\boldsymbol{k}}_2)\cdot \boldsymbol{X}_2}$$

$$= \frac{N_c^2}{N^2} \sum_{\tilde{\boldsymbol{k}}_1, \tilde{\boldsymbol{k}}_2} G(\boldsymbol{X}_1, \boldsymbol{X}_2, \tilde{\boldsymbol{k}}_1, \tilde{\boldsymbol{k}}_2), \tag{15.4}$$

or in matrix notation for the cluster sites $\boldsymbol{X}_1$ and $\boldsymbol{X}_2$

$$\hat{\bar{G}} = \frac{N_c}{N} \sum_{\tilde{\boldsymbol{k}}} \hat{G}(\tilde{\boldsymbol{k}}), \tag{15.5}$$

since $\hat{\bar{G}}$ can be chosen diagonal in $\tilde{\boldsymbol{k}}_1, \tilde{\boldsymbol{k}}_2$. Similarly each interaction line is replaced by

$$\bar{V}(\boldsymbol{X}_1, \boldsymbol{X}_2; \tilde{\boldsymbol{x}} = 0)$$

$$= \frac{1}{N^2} \sum_{\substack{K_1, K_2 \\ \tilde{k}_1, \tilde{k}_2}} e^{i(\boldsymbol{K}_1+\tilde{\boldsymbol{k}}_1)\cdot \boldsymbol{X}_1} V(\boldsymbol{K}_1, \boldsymbol{K}_2; \tilde{\boldsymbol{k}}_1, \tilde{\boldsymbol{k}}_2) e^{-i(\boldsymbol{K}_2+\tilde{\boldsymbol{k}}_2)\cdot \boldsymbol{X}_2}$$

$$= \frac{N_c^2}{N^2} \sum_{\tilde{\boldsymbol{k}}_1, \tilde{\boldsymbol{k}}_2} V(\boldsymbol{X}_1, \boldsymbol{X}_2, \tilde{\boldsymbol{k}}_1, \tilde{\boldsymbol{k}}_2), \tag{15.6}$$

or in matrix notation for the cluster sites $\boldsymbol{X}_1$ and $\boldsymbol{X}_2$

$$\hat{\bar{V}} = \frac{N_c}{N} \sum_{\tilde{\boldsymbol{k}}} \hat{V}(\tilde{\boldsymbol{k}}). \tag{15.7}$$

The summations of the cluster sites $\boldsymbol{X}$ within each diagram remain to be performed. $\hat{\bar{G}}$ and $\hat{\bar{V}}$ are propagators which are truncated outside the cluster. I.e., if the interaction V is non-local, $\hat{\bar{V}}$ will include only interactions within, but not between, clusters. Thus the inclusion of the phase factors $e^{i\tilde{\boldsymbol{k}}\cdot \boldsymbol{X}}$ in the MCA Laue-function (15.2) leads directly to a cluster approach formulated in real space that violates translational invariance. Therefore the Green function and interaction are functions of two cluster momenta $\boldsymbol{K}_1$, $\boldsymbol{K}_2$ or two sites $\boldsymbol{X}_1$, $\boldsymbol{X}_2$ respectively.

If we apply the DCA Laue function (15.3), Green function legs in Φ are replaced by the DCA coarse grained Green function

$$\bar{G}(\boldsymbol{K}) = \frac{N_c}{N} \sum_{\tilde{\boldsymbol{k}}} G(\boldsymbol{K}, \tilde{\boldsymbol{k}}), \tag{15.8}$$

since Green functions can be freely summed over the $\tilde{\boldsymbol{k}}$ vectors within a cell about the cluster momentum $\boldsymbol{K}$. Similarly, the interactions are replaced by the DCA coarse grained interaction

$$\bar{V}(\boldsymbol{K}) = \frac{N_c}{N} \sum_{\tilde{\boldsymbol{k}}} V(\boldsymbol{K}, \tilde{\boldsymbol{k}}). \tag{15.9}$$

As with the MCA, the effect of coarse-graining the interaction is to reduce the effect of non-local interactions to within the cluster. The resulting compact graphs are functionals of the coarse grained Green function $\bar{G}(\boldsymbol{K})$ and interaction $\bar{V}(\boldsymbol{K})$, and thus depend on the cluster momenta $\boldsymbol{K}$ only. For example, when $N_c = 1$, only the local part of the interaction survives the coarse graining. As with the MCA, within the DCA it is important that *both* the interaction and the Green function are coarse-grained [10]. In calculations where a non-local interaction is not coarse-grained, poor results are obtained [13].

15.2.2 From the Cluster to the Lattice

To establish a connection between the cluster and the lattice we minimize the lattice free energy

$$F = -k_B T \left(\Phi - \mathrm{tr}\left[\boldsymbol{\Sigma G}\right] + \mathrm{tr} \ln \left[\boldsymbol{G}\right]\right) , \tag{15.10}$$

where Φ is the generating functional composed of all closed compact (single-particle irreducible) graphs, $\boldsymbol{\Sigma}$ is the lattice self-energy and $\boldsymbol{G}$ is the full lattice Green function. The trace indicates summation over frequency, momentum and spin. As discussed in many-body texts [14], the additional free energy due to an interaction may be described by a sum over all closed connected graphs. These graphs may be further separated into compact and non-compact graphs. The compact graphs, which comprise the generating functional Φ, consist of the sum over all skeletal graphs (those with no internal parts representing corrections to the single-particle Green function). The remaining graphs comprise the non-compact part of the free energy. In the infinite-dimensional limit, Φ consists of only local graphs, with non-local corrections of order $1/D$. However, for the non-compact parts of the free energy, non-local corrections over arbitrary lengths are of order one, so the local approximation applies only to Φ.

To see this, consider the simplest non-local corrections to non-compact and compact parts of the free energy of a Hubbard-like model, illustrated in Fig. 15.2. Here the upper (lower) circle is a set of graphs composed of intrasite propagators restricted to site n (the origin). Consider all such non-local corrections on the shell of sites which are n mutually orthogonal unit translations from the origin. In the limit of high dimensions, there are $2^n D!/((D-n)!n!) \sim \mathcal{O}(D^n)$ such sites. Since as $D \to \infty$, $G(r) \sim D^{-r/2}$ [2], the legs on the compact correction contribute a factor $\mathcal{O}(D^{-2n})$ whereas those on the non-compact correction contribute $\mathcal{O}(D^{-n})$. Therefore the compact non-local correction falls as D^{-n} and is very short-ranged; whereas, the non-compact correction remains of order one, regardless of how far site n is from the origin [15]. As we will see below, the essential approximation of the DCA and the MCA is to use the cluster propagators, which are accurate only for short distances, to construct various diagrammatic insertions. In high dimensions, or in finite

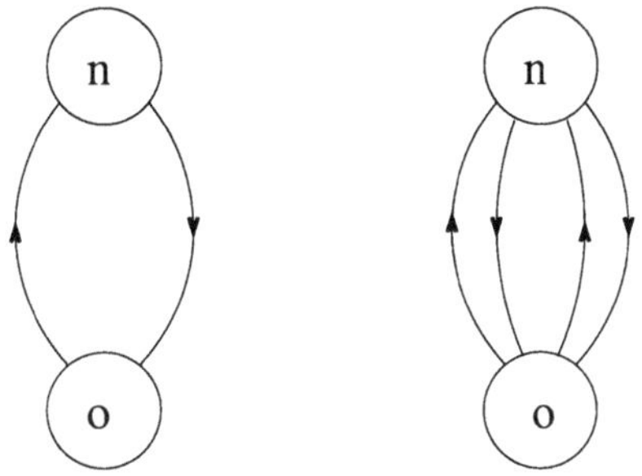

Fig. 15.2. Non-compact (left) and compact (right) non-local corrections to the free energy functional. Here the upper (lower) circle is meant to represent a set of graphs which are closed except for the external lines shown, and restricted to site n (the origin)

dimensions when the Green functions fall exponentially with distance, this is a good approximation for the compact graphs which comprise Φ, but a poor approximation for the non-compact graphs [15].

Thus, we will approximate the generating functional Φ with its cluster counterpart Φ_c by replacing the Laue function with either Δ_{DC} or Δ_{MC}, but this approximation will not be used in the parts of the free energy coming from non-compact graphs. The free energy then reads

$$F = -k_B T \left(\Phi_c - \mathrm{tr}\left[\boldsymbol{\Sigma G}\right] + \mathrm{tr} \ln \left[\boldsymbol{G}\right] \right) . \tag{15.11}$$

F is stationary with respect to $\boldsymbol{G}$ when $\frac{\delta F}{\delta G} = 0$. This happens for the MCA if we estimate the lattice self energy as

$$\Sigma(\boldsymbol{K}_1, \boldsymbol{K}_2; \tilde{\boldsymbol{k}}_1, \tilde{\boldsymbol{k}}_2) = \sum_{\boldsymbol{X}_1, \boldsymbol{X}_2} e^{-i(\boldsymbol{K}_1 + \tilde{\boldsymbol{k}}_1)\cdot \boldsymbol{X}_1} \Sigma_{\mathrm{MC}}(\boldsymbol{X}_1, \boldsymbol{X}_2) e^{i(\boldsymbol{K}_2 + \tilde{\boldsymbol{k}}_2)\cdot \boldsymbol{X}_2} . \tag{15.12}$$

Thus, the corresponding lattice single-particle propagator reads in matrix notation

$$\hat{G}(\tilde{\boldsymbol{k}}, z) = \left[zI - \hat{\epsilon}(\tilde{\boldsymbol{k}}) - \hat{\Sigma}_{MC}(z) \right]^{-1} , \tag{15.13}$$

where the dispersion $\hat{\epsilon}(\tilde{\boldsymbol{k}})$ and self-energy $\hat{\Sigma}_{\mathrm{MC}}(z)$ are matrices in cluster real space with

$$[\hat{\epsilon}(\tilde{\boldsymbol{k}})]_{\boldsymbol{X}_1 \boldsymbol{X}_2} = \epsilon(\boldsymbol{X}_1 - \boldsymbol{X}_2, \tilde{\boldsymbol{k}}) = \frac{1}{N_c} \sum_{\boldsymbol{K}} e^{i(\boldsymbol{K} + \tilde{\boldsymbol{k}})\cdot(\boldsymbol{X}_1 - \boldsymbol{X}_2)} \tag{15.14}$$

being the intracluster Fourier transform of the dispersion. For the DCA, $\Sigma(\boldsymbol{k}) = \Sigma_{\mathrm{DC}}(\boldsymbol{K})$ is the proper approximation for the lattice self energy corresponding to Φ_{DC}. The corresponding lattice single-particle propagator is then given by

$$G(\boldsymbol{K}, \tilde{\boldsymbol{k}}; z) = \frac{1}{z - \epsilon_{\boldsymbol{K} + \tilde{\boldsymbol{k}}} - \Sigma_{\mathrm{DC}}(\boldsymbol{K}, z)} . \tag{15.15}$$

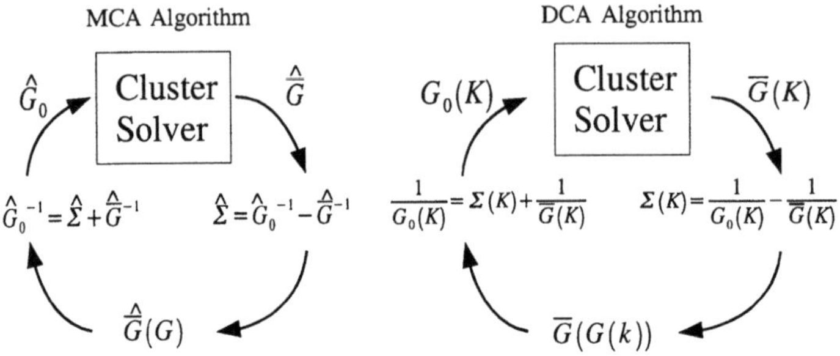

Fig. 15.3. The MCA and DCA algorithms. The two differ mainly in that in the DCA, the Green functions are diagonal in $\boldsymbol{K}$, while in the MCA, they are matrices in the cluster coordinates $\boldsymbol{X}$

Both the MCA and DCA are optimized when we equate the lattice and cluster self energies. A similar relation holds for two-particle quantities. Thus, with few exceptions [16], only the irreducible quantities on the cluster and lattice correspond one-to-one.

The MCA (DCA) algorithm, illustrated in Fig. 15.3, follows directly: We first make an initial guess for the cluster self-energy matrix Σ. This is used with (15.5) and (15.13) [(15.8) and (15.15)] to calculate the coarse-grained Green function $\bar{G}$. The cluster excluded Green function $\hat{G}_0 = [\hat{\bar{G}}^{-1} + \hat{\Sigma}_{MC}]^{-1}$ ($G_0(\boldsymbol{K}) = [\bar{G}(\boldsymbol{K})^{-1} + \Sigma_{\rm DC}(\boldsymbol{K})]^{-1}$) is defined to avoid overcounting self energy corrections on the cluster. It is used to compute a new estimate for the cluster self-energy which is used to reinitialize the process.

Once convergence is reached, the irreducible quantities on the cluster may be used to calculate the corresponding lattice quantities. For example, the cluster self energy and irreducible vertex functions may be used in the Dyson and Bethe–Salpeter equations to calculate the Green function and susceptibilities. In order to obtain a smooth DCA self energy for the calculation of the band structure and spectra, it may be interpolated into the Brillouin zone of the lattice (however, such interpolation should be avoided during the self-consistency loop, as it may lead to causality violations). The optimal MCA self energy, (15.12) is a function of two momenta since the translational invariance of the lattice is violated. Kotliar et al. have introduced a cluster averaging scheme to obtain a self energy as a function of one momentum which may be employed after convergence is obtained [12].

15.2.3 The Small Parameter, $\hat{\Gamma}$: The Coupling Between the Cluster and Its Host

In order to compare the character of the two different cluster approaches as a function of the cluster size N_c it is instructive to rewrite the corresponding coarse grained Green-functions (15.5) and (15.8) to suitable forms by making use of the independence of the self-energy Σ on the integration variable $\tilde{\boldsymbol{k}}$. For the MCA coarse grained Green function we find

$$\hat{\bar{G}}(z) = \left[zI - \hat{\epsilon}_o - \hat{\Sigma}_{\mathrm{MC}}(z) - \hat{\Gamma}_{\mathrm{MC}}(z)\right]^{-1}, \tag{15.16}$$

with the "cluster-local" energy $\hat{\epsilon}_o = N_c/N \sum_{\tilde{\boldsymbol{k}}} \hat{\epsilon}(\tilde{\boldsymbol{k}})$. For the DCA we obtain a similar expression

$$\bar{G}(\boldsymbol{K}, z) = \left[z - \bar{\epsilon}_{\boldsymbol{K}} - \Sigma_{\mathrm{DC}}(\boldsymbol{K}, z) - \Gamma_{\mathrm{DC}}(\boldsymbol{K}, z)\right]^{-1}, \tag{15.17}$$

with the coarse grained average $\bar{\epsilon}_{\boldsymbol{K}} = N_c/N \sum_{\tilde{\boldsymbol{k}}} \epsilon(\boldsymbol{K}, \tilde{\boldsymbol{k}})$. The hybridization functions $\hat{\Gamma}_{MC/DC}(z)$ describe the coupling of the cluster to the mean-field representing the remainder of the system.

The behavior of Γ for large N_c is important. For the MCA, Γ averaged over the cluster sites and frequency

$$\bar{\Gamma}_{MC} = \frac{1}{N_c} \sum_{\boldsymbol{X}_1, \boldsymbol{X}_2} \Gamma_{MC}(\boldsymbol{X}_1, \boldsymbol{X}_2) \sim \mathcal{O}\left(\frac{2D}{L_c}\right), \tag{15.18}$$

where $L_c = N_c^{1/D}$ is the linear cluster size. A detailed derivation of this form is presented in the appendix. However, since in the MCA the cluster is defined in real space with open boundary conditions, this form is evident since only the sites on the surface $\propto 2D \cdot L_c^{D-1}$ of the cluster couple to the effective medium and $N_c = L_c^D$. For the DCA we show in the appendix that $\Gamma(\boldsymbol{K}) \sim \mathcal{O}(1/N_c^{2/D})$ (see also [11]) so that we obtain for the average hybridization of the DCA cluster to the effective medium

$$\bar{\Gamma}_{DC} = \frac{1}{N_c} \sum_{\boldsymbol{K}} \Gamma_{DC}(\boldsymbol{K}) \sim \mathcal{O}\left(\frac{1}{L_c^2}\right). \tag{15.19}$$

The DCA coarse graining results in a cluster in $\boldsymbol{K}$-space; thus, the corresponding real space cluster has periodic boundary conditions, and each site in the cluster has the same hybridization strength $\bar{\Gamma}$ with the host.

As shown in the appendix, the average hybridization strength $\bar{\Gamma}$ acts as the small parameter in both the MCA and the DCA. Thus the MCA (DCA) is an approximation with corrections of order $\bar{\Gamma} \sim \mathcal{O}(1/L_c)$ ($\sim \mathcal{O}(1/L_c^2)$).

15.3 Numerical Results

To illustrate the differences in convergence with cluster size N_c we performed MCA and DCA simulations for the symmetric one-dimensional (1D) Falicov–Kimball model (FKM). At half filling the FKM Hamiltonian reads

$$H = -t \sum_i (d_i^\dagger d_{i+1} + h.c.) + U \sum_i (n_i^d - 1/2)(n_i^f - 1/2), \tag{15.20}$$

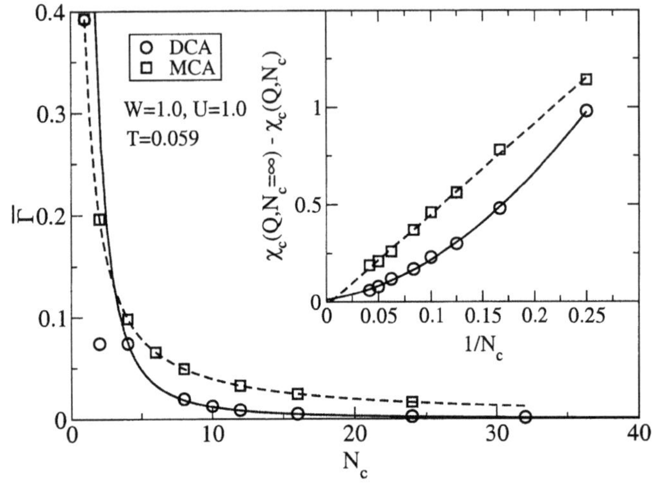

Fig. 15.4. The average integrated hybridization strengths $\bar{\Gamma}$ of the MCA (squares) and DCA (circles) versus the cluster size N_c when $\beta = 17$ and $U = W = 1$ in the symmetric model. The solid and dashed lines represent the fits $1.1946/N_c^2$ and $0.3361/N_c$ respectively. Inset: Convergence of the cluster charge susceptibility for $Q = \pi$. The solid and dashed lines are quadratic and linear fits, respectively

with the number operators $n_i^d = d_i^\dagger d_i$ and $n_i^f = f_i^\dagger f_i$ and the Coulomb repulsion U between d and f electrons residing on the same site. The FKM can be considered as a simplified Hubbard model with only one spin-species (d) being allowed to hop. However it still shows a complex phase diagram including a Mott gap for large U and half filling, an Ising-like charge ordering with the corresponding transition temperature T_c being zero in 1D, and phase separation in all dimensions. The bare dispersion (in 1D) $\epsilon_k = 2t \cos k$; thus for $t = 1/4$ the bandwidth $W = 1$ which we use as unit of energy. To simulate the effective cluster models of the MCA an the DCA we use a quantum Monte Carlo (QMC) approach described in [10].

To check the scaling relations (15.18) and (15.19), we show in Fig. 15.4 the average hybridization functions $\bar{\Gamma}_{\mathrm{MC}}$ and $\bar{\Gamma}_{\mathrm{DC}}$ for the MCA and DCA respectively at the inverse temperature $\beta = 17$ for $U = W = 1$. For $N_c = 1$ both approaches are equivalent to the DMFA and thus $\bar{\Gamma}_{\mathrm{MC}} = \bar{\Gamma}_{\mathrm{DC}}$. For increasing N_c $\bar{\Gamma}_{\mathrm{MC}}$ can be fitted by $0.3361/N_c$ and $\bar{\Gamma}_{\mathrm{DC}}$ by $1.1946/N_c^2$ when $N_c > 2$. Cluster quantities, such as the self energy and cluster susceptibilities, are expected to converge with increasing N_c like $\bar{\Gamma}$. This is illustrated in the inset for the staggered ($Q = \pi$) charge susceptibility $\chi_c(\mathbf{Q})$ of the cluster.

Since only the compact parts represented by Φ of the lattice free energy (15.10) are coarse-grained, this scaling is expected to break down when lattice quantities, such as the lattice charge susceptibility, are calculated. The susceptibility of the cluster $\chi_c(Q)$ cannot diverge for any finite N_c; whereas the lattice $\chi(Q)$ diverges at the transition temperature T_c to the charge or-

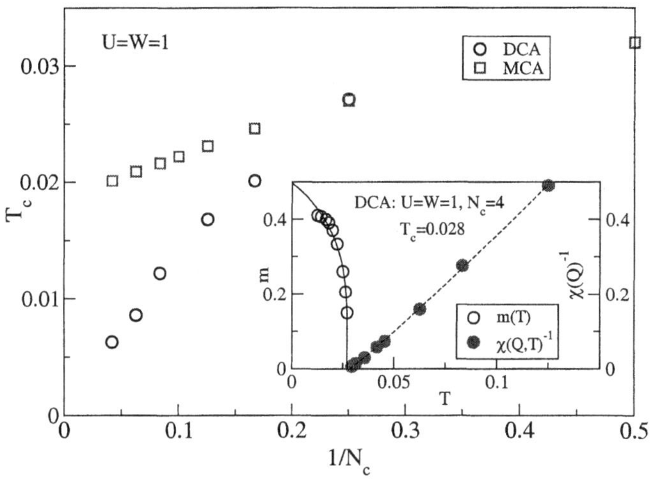

Fig. 15.5. The transition temperature T_c for the DCA (circles) and MCA (squares) when $U = W = 1$ versus the cluster size N_c. For all values of N_c the DCA prediction is closer to the exact result ($T_c = 0$). Inset: Order parameter $m(T)$ and inverse charge susceptibility $\chi(Q)^{-1}$ versus temperature. The solid (dashed) line represents a fit to the functions $m(T) \propto (T_c - T)^\beta$ with $\beta = 0.245$ ($\chi(T) \propto (T - T_c)^{-\gamma}$ with $\gamma = 1.07$)

dered phase. Note that the residual mean-field character of both methods can result in finite transition temperatures $T_c > 0$ for finite $N_c < \infty$. However as N_c increases, this residual mean field character decreases gradually and thus increased fluctuations should drive the solution to the exact result $T_c = 0$.

In the DCA [10], $\chi(Q)$ is calculated by first extracting the corresponding vertex function from the cluster simulation. This is then used in a Bethe–Salpeter equation to calculate $\chi(Q)$. T_c is calculated by extrapolating $\chi(Q)^{-1}$ to zero using the function $\chi(Q)^{-1} \propto (T - T_c)^\gamma$ (see inset to Fig. 15.5). This procedure is difficult, if not impossible, in the MCA due to the lack of translational invariance. Here, we calculate the order parameter $m(T) = 1/N_c \sum_i (-1)^i \langle n_i^d \rangle$ in the symmetry broken phase. T_c is then obtained from extrapolating $m(T)$ to zero using the function $m(T) \propto (T_c - T)^\beta$. For the DCA this extrapolation is shown by the solid line in the inset to Fig. 15.5 for $N_c = 4$. The values for T_c obtained from the calculation in the symmetry broken phase and in the unbroken phase must agree, since as we have shown above, both the DCA and MCA are Φ-derivable. This is illustrated in Fig. 15.5 for the DCA.

A comparison of the DCA and MCA estimate of T_c is presented in Fig. 15.5. T_c obtained from MCA (squares) is larger than T_c obtained from DCA (circles). Moreover we find that the DCA result seems to scale to zero almost linearly in $1/N_c$ (for large enough N_c), whereas the MCA does not show any scaling form and in fact seems to tend to a finite value for T_c as $N_c \to \infty$.

This striking difference of the two methods can be attributed to the different boundary conditions. The open boundary conditions of the MCA cluster result in a large surface contribution so that $\bar{\Gamma}_{\mathrm{MC}} > \bar{\Gamma}_{\mathrm{DC}}$. This engenders pronounced mean field behavior that stabilizes the finite temperature transition for the cluster sizes treated here. For larger clusters we expect the bulk contribution to the MCA free energy to dominate so that T_c should fall to zero.

Complementary results are found in simulations of *finite-sized* systems. In general, systems with open boundary conditions are expected to have a surface contribution in the free energy of order $\mathcal{O}(1/L_c)$ [17]. This term is absent in systems with periodic boundary conditions. As a result, simulations of finite-sized systems with periodic boundary conditions converge much more quickly than those with open boundary conditions [18].

Thus far, we have shown that the DCA converges more quickly than the MCA for critical properties and for extended cluster quantities (e.g. the cluster susceptibility). This is due to the differences in the boundary conditions, and the coupling to the mean-field. Whereas each site in the DCA experiences the same coupling to the mean-field host, in the MCA only the sites on the boundary of the cluster couple to the host. Provided that the system is far from a transition, the sites in the center of the cluster couple to the mean-field only through propagators which fall exponentially with distance. Thus, one might expect that local results, such as the single-particle density of states, might converge more quickly within the MCA provided that they are measured on these central sites. This is illustrated in Fig. 15.6 where we plot the single-particle density of states calculated with the DCA and the MCA on the two central sites.

15.4 Summary

By defining appropriate Laue functions, we provide microscopic diagrammatic derivations of the MCA and DCA. We show that they are Φ-derivable, and that the lattice free energy is optimized by equating the irreducible quantities on the lattice to those on the cluster. The MCA maps the lattice to a cluster with open boundaries and consequently, the cluster violates translational invariance. In contrast, the DCA cluster has periodic boundary conditions, and therefore preserves the translational invariance of the lattice.

These differences in the boundary conditions translate directly to different asymptotic behaviors for large clusters N_c. As we find analytically as well as numerically, the surface contributions in the MCA lead to an average hybridization $\bar{\Gamma}$ of the cluster to the mean field that scales like $1/L_c$ as compared to the $1/L_c^2$ scaling of the DCA. Since $\bar{\Gamma}$ acts as the small parameter for these approximation schemes, the DCA converges much more quickly than the MCA. These effects are more pronounced near a transition, where the large surface contribution of the MCA stabilizes the mean-field character of

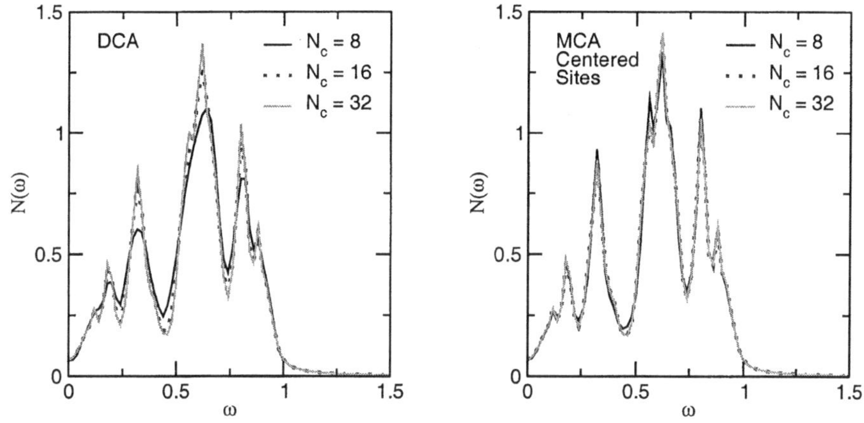

Fig. 15.6. Comparison of the single-particle density of states (DOS) calculated with the DCA (left) and the MCA (right) on the two central sites for various cluster sizes N_c when $\beta = 10$ and $U = W = 1$ (since the DOS is symmetric around $\omega = 0$, only the $\omega \geq 0$ part is shown). Local quantities, such as the DOS, converge more quickly when calculated on the central sites of the MCA cluster than the cluster averaged DCA result

the transition. Consequently, the DCA result for the transition temperature T_c of the 1D symmetric FKM model scales almost like $1/N_c$ to the exact result $T_c = 0$, whereas the MCA result converges very slowly. The boundary conditions also differ in that only the MCA sites at the surface of the cluster couple to the mean field; whereas, all DCA cluster sites have an equal coupling to the mean field host. As a result local quantities, such as the density of states, when measured on the central sites of the MCA cluster converge more quickly than than corresponding DCA results.

Thus, for critical properties and extended cluster quantities, the DCA converges far more quickly than the MCA; whereas for local quantities which may be measured at the central sites of the MCA cluster, the MCA converges more quickly. Since the origin of these differences lies in the different boundary conditions we expect them to hold generally for any model of electrons moving on a lattice.

Acknowledgements

We acknowledge useful conversations with N. Blümer, A. Gonis, M. Hettler, H.R. Krishnamurthy, D.P. Landau, T. Pruschke, W. Shelton, and A. Voigt. This work was supported by NSF grants DMR-0073308 and 0113574. This research was supported in part by NSF cooperative agreement ACI-9619020 through computing resources provided by the National Partnership for Advanced Computational Infrastructure at the Pittsburgh Supercomputer Center.

Appendix

To differentiate the MCA from the DCA we find expressions for the corresponding host-functions Γ. To this end we split the hopping integral t into an inter- and intracluster part. For the MCA the intercluster hopping $[\hat{T}_{\mathrm{MC}}]_{\boldsymbol{X}_1,\boldsymbol{X}_2} = T_{MC}(\boldsymbol{X}_1, \boldsymbol{X}_2)$ is defined as $\hat{T}_{\mathrm{MC}}(\tilde{\boldsymbol{k}}) = \hat{\epsilon}(\tilde{\boldsymbol{k}}) - \hat{\epsilon}_o$ and reads in real space

$$\hat{T}_{\mathrm{MC}}(\tilde{\boldsymbol{x}}) = \hat{t}(\tilde{\boldsymbol{x}}) - \hat{\epsilon}_o \delta(\tilde{\boldsymbol{x}}) . \tag{15.21}$$

Since $\hat{\epsilon}_o = \hat{t}(\tilde{\boldsymbol{x}} = 0)$, the intercluster hopping matrix $\hat{T}_{\mathrm{MC}}(\tilde{\boldsymbol{x}})$ is only finite for $\tilde{\boldsymbol{x}} \neq 0$ as expected. Thus, $\hat{T}_{\mathrm{MC}}(\tilde{\boldsymbol{x}})$ has non-vanishing matrix-elements only for sites on the boundary of the cluster. For the DCA the intercluster hopping integral is analogously defined as $T_{\mathrm{DC}}(\boldsymbol{K}, \tilde{\boldsymbol{k}}) = \epsilon(\boldsymbol{K}, \tilde{\boldsymbol{k}}) - \bar{\epsilon}_{\boldsymbol{K}}$ and can be written in real space as

$$\hat{T}_{\mathrm{DC}}(\tilde{\boldsymbol{x}}) = \hat{t}(\tilde{\boldsymbol{x}}) - \hat{\bar{t}}\hat{\delta}^c(\tilde{\boldsymbol{x}}) , \tag{15.22}$$

where $[\hat{\bar{t}}]_{\boldsymbol{X}_1 \boldsymbol{X}_2} = \bar{t}(\boldsymbol{X}_1 - \boldsymbol{X}_2) = 1/N_c \sum_{\boldsymbol{K}} e^{i\boldsymbol{K}\cdot(\boldsymbol{X}_1 - \boldsymbol{X}_2)} \bar{\epsilon}_{\boldsymbol{K}}$ and the cluster delta-like function $[\hat{\delta}^c(\tilde{\boldsymbol{x}})]_{\boldsymbol{X}_1 \boldsymbol{X}_2} = \delta^c(\boldsymbol{X}_1 - \boldsymbol{X}_2, \tilde{\boldsymbol{x}}) = N_c/N \sum_{\tilde{\boldsymbol{k}}} e^{i\tilde{\boldsymbol{k}}\cdot(\boldsymbol{X}_1 - \boldsymbol{X}_2 + \tilde{\boldsymbol{x}})}$. We have discussed elsewhere that $\delta^c(\boldsymbol{X}_1 - \boldsymbol{X}_2, \tilde{\boldsymbol{x}}) \approx 1$ for $\tilde{\boldsymbol{x}} = 0$ and falls off rapidly for finite $\tilde{\boldsymbol{x}}$. It is important to note that $\bar{t}(\boldsymbol{X}_1 - \boldsymbol{X}_2) = t(\boldsymbol{X}_1 - \boldsymbol{X}_2, \tilde{\boldsymbol{x}} = 0)$ for $N_c = N$, but $|\bar{t}(\boldsymbol{X}_1 - \boldsymbol{X}_2)| < |t(\boldsymbol{X}_1 - \boldsymbol{X}_2, \tilde{\boldsymbol{x}} = 0)|$ for $N_c < N$. It follows that the DCA intercluster hopping $T_{DC}(\boldsymbol{X}_1 - \boldsymbol{X}_2, \tilde{\boldsymbol{x}})$ for $\tilde{\boldsymbol{x}} = 0$, i.e. between sites belonging to the same cluster, is even finite when $N_c < N$; however it is strongly reduced compared to $\tilde{\boldsymbol{x}} \neq 0$. This is a consequence of the periodicity of the DCA cluster. Therefore we find that the effective model of the DCA cannot be thought of N/N_c clusters composing the original lattice, but we have to rather think of it as a renormalized model in real space.

Now we proceed with finding expressions for the hybridization functions Γ defined in (15.16) and (15.17) for the MCA and DCA. We start with the MCA. To this end we define a Green function matrix

$$\hat{g}(z) = \left[zI - \hat{\epsilon} - \hat{\Sigma}_{\mathrm{MC}}(z) \right]^{-1} , \tag{15.23}$$

that is localized on the impurity cluster. With this definition (15.5) reads

$$\hat{\bar{G}} = \frac{N_c}{N} \sum_{\tilde{\boldsymbol{k}}} \hat{G}(\tilde{\boldsymbol{k}}) = \left[\hat{g}^{-1} - \hat{\Gamma}_{\mathrm{MC}} \right]^{-1} , \tag{15.24}$$

and (15.13) can be rewritten to

$$\hat{G}(\tilde{\boldsymbol{k}}) = \hat{g} + \hat{g}\hat{T}_{\mathrm{MC}}(\tilde{\boldsymbol{k}})\hat{G}(\tilde{\boldsymbol{k}}) . \tag{15.25}$$

After inserting (15.25) into (15.24) and using $\sum_{\tilde{\boldsymbol{k}}} \hat{T}_{\mathrm{MC}}(\tilde{\boldsymbol{k}}) = 0$ we obtain after some algebraic transformations for the MCA-hybridization matrix

$$\hat{\Gamma}_{\mathrm{MC}} = \left[I + \frac{N_c}{N} \sum_{\tilde{\boldsymbol{k}}} \hat{T}_{\mathrm{MC}}(\tilde{\boldsymbol{k}}) \hat{G}(\tilde{\boldsymbol{k}}) \right]^{-1} \times \frac{N_c}{N} \sum_{\tilde{\boldsymbol{k}}} \hat{T}_{\mathrm{MC}}(\tilde{\boldsymbol{k}}) \hat{G}(\tilde{\boldsymbol{k}}) \hat{T}_{\mathrm{MC}}(\tilde{\boldsymbol{k}}) . \tag{15.26}$$

Since, as we pointed out above, $\hat{T}$ has non-zero matrix-elements only between sites on the surface of the impurity cluster, $\bar{\Gamma}$ is finite only for sites on the surface but vanishes for sites inside the cluster. Thus, the average hybridization strength of the MCA cluster per site

$$\bar{\Gamma}_{\mathrm{MC}} = \frac{1}{N_c} \sum_{\boldsymbol{X}_1, \boldsymbol{X}_2} \Gamma_{\mathrm{MC}}(\boldsymbol{X}_1, \boldsymbol{X}_2) \sim \mathcal{O}\left(\frac{2D}{L_c}\right), \tag{15.27}$$

since only the sites on the surface $\propto 2D \cdot L_c^{D-1}$ of the cluster couple to the effective medium and $N_c = L_c^D$.

For the DCA we can follow the steps presented above for the MCA since $\Gamma_{\mathrm{DC}}(\boldsymbol{K})$ can be considered a diagonal matrix in $\boldsymbol{K}$. We obtain for the DCA hybridization function

$$\Gamma_{\mathrm{DC}}(\boldsymbol{K}) = \frac{\frac{N_c}{N} \sum_{\tilde{\boldsymbol{k}}} T_{\mathrm{DC}}^2(\boldsymbol{K}, \tilde{\boldsymbol{k}}) G(\boldsymbol{K}, \tilde{\boldsymbol{k}})}{1 + \frac{N_c}{N} \sum_{\tilde{\boldsymbol{k}}} T_{\mathrm{DC}}(\boldsymbol{K}, \tilde{\boldsymbol{k}}) G(\boldsymbol{K}, \tilde{\boldsymbol{k}})}. \tag{15.28}$$

It should be stressed that due to the periodicity of the DCA cluster every site of the impurity cluster couples to the effective medium. The momenta $\tilde{\boldsymbol{k}}$ are restricted to a DCA coarse graining cell and therefore maximal of the order $\mathcal{O}(\Delta k)$, where $\Delta k \propto 1/L$. Since $\epsilon(\boldsymbol{K}) - \bar{\epsilon}_{\boldsymbol{K}} \sim \mathcal{O}((\Delta k)^2)$, we find by performing a Taylor series expansion of $T_{\mathrm{DC}}(\boldsymbol{K}, \tilde{\boldsymbol{k}})$ around $\epsilon(\boldsymbol{K})$ that $T_{\mathrm{DC}}(\boldsymbol{K}, \tilde{\boldsymbol{k}}) \sim \mathcal{O}(1/L_c)$. For the overall hybridization of the DCA cluster to the effective medium we thus obtain

$$\bar{\Gamma}_{\mathrm{DC}} = \frac{1}{N_c} \sum_{\boldsymbol{K}} \Gamma_{\mathrm{DC}}(\boldsymbol{K}) \sim \mathcal{O}\left(\frac{1}{L_c^2}\right). \tag{15.29}$$

In both the DCA and the MCA, the average hybridization strength acts as the small parameter. The approximation performed by the DCA (MCA) is to replace the lattice Green function $G(\boldsymbol{K}, \tilde{\boldsymbol{k}}, z) = [z - \epsilon_{\boldsymbol{K}+\tilde{\boldsymbol{k}}} - \Sigma(\boldsymbol{K}, \tilde{\boldsymbol{k}}, z)]^{-1}$ ($\hat{G}(\tilde{\boldsymbol{k}}) = [zI - \hat{\epsilon}(\tilde{\boldsymbol{k}}) - \hat{\Sigma}(\tilde{\boldsymbol{k}}, z)]^{-1}$) by its coarse grained quantity $\bar{G}(\boldsymbol{K})$ ($\hat{\bar{G}}$) in diagrams for the generating functional Φ.

According to (15.17) the coarse grained Green function of the DCA can be expressed as $\bar{G}(\boldsymbol{K}, z) = [z - \bar{\epsilon}_{\boldsymbol{K}} - \Sigma_{\mathrm{DC}}(\boldsymbol{K}, z) - \Gamma_{\mathrm{DC}}(\boldsymbol{K}, z)]^{-1}$. For the time being, assume that $\Sigma_{\mathrm{DC}}(\boldsymbol{K}, z)$ has corrections of the same, or higher, order in $1/L_c$ as the average hybridization strength. Both the self energy and $\epsilon_{\boldsymbol{K}+\tilde{\boldsymbol{k}}} = \bar{\epsilon}_{\boldsymbol{K}} + \mathcal{O}(1/L_c)$ with the leading order corrections being linear in $\tilde{\boldsymbol{k}}$. Since furthermore $\Gamma_{\mathrm{DC}}(\boldsymbol{K}) \sim \mathcal{O}(1/L_c^2)$, $G(\boldsymbol{K}, \tilde{\boldsymbol{k}}) \approx \bar{G}(\boldsymbol{K}) + \mathcal{O}(1/L_c)$. The diagrams for Φ however are summed over $\tilde{\boldsymbol{k}}$, so that the terms $\sim \mathcal{O}(1/L_c)$ coming from $\epsilon_{\boldsymbol{K}+\tilde{\boldsymbol{k}}}$ and similar terms from the self energy vanish and only the terms $\sim \mathcal{O}(1/L_c^2)$, or higher, survive. Thus we find for the DCA that $\Phi \approx \Phi_{\mathrm{DC}} + \mathcal{O}(1/L_c^2)$. The corresponding estimate of $\Sigma_{\mathrm{DC}}(\boldsymbol{K}, z)$ will also have corrections $\mathcal{O}(1/L_c^2)$, confirming the assumption above.

According to (15.16), the coarse grained propagator of the MCA can be written as $\hat{\bar{G}}(z) = [zI - \hat{\epsilon}_o - \hat{\Sigma}_{\rm MC}(z) - \hat{\Gamma}_{\rm MC}(z)]^{-1}$. If we assume that $\hat{\Sigma}_{\rm MC}$ converges with L_c as fast or faster than $1/L_c$ we see with the L_c-dependence of $\hat{\Gamma}_{\rm MC}$ in (15.27) that $\hat{G}(\tilde{\boldsymbol{k}}) \approx \hat{\bar{G}} + \mathcal{O}(1/L_c)$ since $\hat{\epsilon}(\tilde{\boldsymbol{k}}) = \hat{\epsilon}_o + \mathcal{O}(1/L_c)$. Thus we obtain for the MCA approximation $\Phi \approx \Phi_{\rm MC} + \mathcal{O}(1/L_c)$ and $\hat{\Sigma}$ will converge with L_c as $\mathcal{O}(1/L_c)$ confirming our assumption.

References

1. D.W. Taylor: Phys. Rev. **156**, 1017 (1967); P. Soven Phys. Rev. **156**, 809 (1967); P.L. Leath and B. Goodman, Phys. Rev. **148**, 968 (1966)
2. W. Metzner and D. Vollhardt: Phys. Rev. Lett. **62**, 324 (1989)
3. E. Müller-Hartmann: Z. Phys. B **74**, 507 (1989)
4. T. Pruschke, M. Jarrell and J.K. Freericks: Adv. in Phys. **42**, 187 (1995)
5. A. Georges, G. Kotliar, W. Krauth and M.J. Rozenberg: Rev. Mod. Phys. **68**, 13 (1996)
6. M. Jarrell and H.R. Krishnamurthy: Phys. Rev. B **63**, 125102 (2001)
7. F. Ducastelle: J. Phys. C. Sol. State. Phys. **7**, 1795 (1974)
8. A. Gonis: *Green functions for ordered and disordered systems* (Studies in Mathematical Physics) Eds. E. van Groesen and E.M. DeJager (North Holland, Amsterdam 1992)
9. M.H. Hettler, A.N. Tahvildar-Zadeh and M. Jarrell: Phys. Rev. B **58**, 7475 (1998)
10. M.H. Hettler, M. Mukherjee, M. Jarrell, and H.R. Krishnamurthy: Phys. Rev. B **61**, 12739 (2000)
11. T. Maier, M. Jarrell, T. Pruschke and J. Keller: Eur. Phys. J. B **13**, 613 (2000)
12. G. Kotliar, S.Y. Savrasov, G. Pálsson and G. Biroli: Phys. Rev. Lett. **87**, 186401 (2001). This method reduces to the MCA when non-overlapping clusters are chosen
13. G. Biroli and G. Kotlia: Phys. Rev. B **65**, 155112 (2002)
14. A.A. Abrikosov, L.P. Gorkov, I.E. Dzyalishinski: *Methods of Quantum Field Theory in Statistical Physics* (Dover, New York 1975)
15. K. Aryanpour, M.H. Hettler and M. Jarrell: Phys. Rev. B **65**, 153102 (2002)
16. In both the DCA and MCA, on-site lattice and cluster quantities correspond one-to-one. In the MCA, the single-particle green functions within the cluster and the lattice also correspond
17. M.E. Fisher and M.N. Barber: Phys. Rev. Lett. **28**, 1516 (1972)
18. D.P. Landau: Phys. Rev. B **13**, 2997 (1976)

Part III

Materials Properties

16 Asymptotic Behavior of a Strain Percolation Model for a Deforming Metal

Y. Shim[1,2], L.E. Levine[2], R. Thomson[2], and D.E. Kramer[2]

[1] Center for Simulation Physics, University of Georgia, Ahtens, GA 30602, USA
[2] Materials Science and Engineering Laboratory, National Institute of Standards and Technology, Gaithersburg, MD 20899, USA

Abstract. In this paper, we present a recent advance in theoretical understanding of a deforming metal, using a strain percolation model which possibly explains spasmodic, fine slip line burst events occurring in the metal. The model addresses how the additional strain nucleated in a cell propagates through a dislocation cell structure, and predicts that near the critical point, the system exhibits critical power-law behavior. It is found that although the model displays long-transient behavior associated with the initial strain in the model, asymptotically critical behavior observed in the system is well explained by standard percolation theory. The long-transient behavior suggests that finite-size effects could be an important factor for the stress-strain relation in the metal. A detailed study reveals that the universal aspects of the model, i.e., the evolution into an initial condition-independent, critical state, arise from collective behavior of a huge number of self-organizing critical cells that develop the minimum or at least marginally stable strain.

16.1 Introduction

Dislocations are line defects generated during a deformation process, and they interact with long-range $1/r$ as well as short-range forces in the presence of a driving force, the external stress. The collective motion of dislocations under stress produces irreversible plastic flow along a slip direction in a deforming metal. Since dislocation movement takes place on some specific slip planes (although movement out of the slip plane is possible by climb), the plastic flow is, by nature, highly anisotropic and inhomogeneous in time and space. The slip plane is normally the plane with the highest density of atoms and the slip direction is the direction in the slip plane in which the atoms are most closely spaced [1]. As a result of such complicated interactions with glide plane constraints, various types of a dislocation structure emerge. But, because of the enormous complexity, metal deformation is poorly understood.

One interesting feature resulted from the deformation is that the slip appears on the surface as long fine slip lines, and bands composed of the fine slip lines, as shown in Fig. 16.1a. These bands are distributed inhomogeneously in the sample and the surface area between the bands is nearly slip free. Miguel

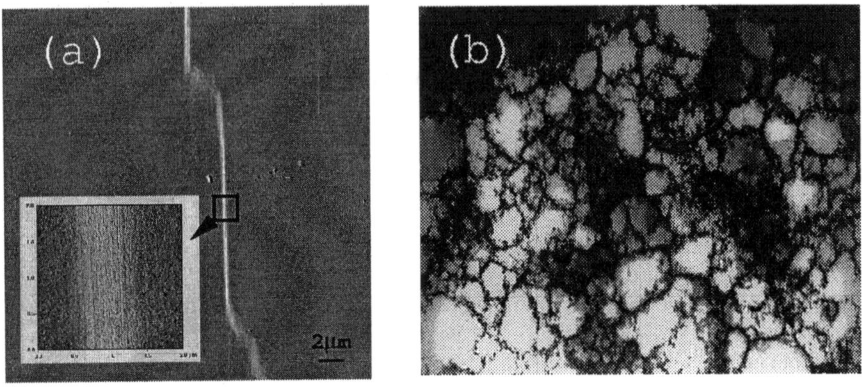

Fig. 16.1. (**a**) AFM micrograph of a deformed Al single crystal, showing a surface slip band. The inset of (**a**) is a magnified image of the region denoted by a box ($2\,\mu m \times 2\,\mu m$), which shows fine slip lines within the band. (**b**) TEM micrograph of a Cu single crystal deformed by uniaxial tension along [100]; after H. Mughrabi et al. [7], showing cell walls (dark regions) and cell interior (bright regions)

et al. have reported a statistical analysis of the acoustic emissions induced by dislocation motion during creep of an ice single crystal, which shows that the distribution of the acoustic energy bursts (which are probably related to slip line burst events) exhibits a power-law decay over several decades [2]. They also suggested that the deformation in ice could be described in the framework of non-equilibrium critical phenomena [3,4]. However, it is not clear whether or not such power-law behavior observed in ice is applicable to a deforming metal.

During plastic deformation, some of the mobile dislocations are trapped and become immobilized. Due to the formation of immobilized structures, further deformation with the transport of mobile dislocations through such structures becomes increasingly difficult. This phenomenon is called *work hardening*. One can view this process as a pinning-depinning transition with the presence of a critical flow stress. Sevillano et al. [5] have explored intrinsic size effects in plasticity near the pinning-depinning transition and discussed a similarity between a fluid invasion of porous media (an invasion percolation problem) and dislocation glide through non-shearable and shearable obstacles. Zapperi and Zaiser have studied roughening of a dislocation line after the depinning transition and showed that the roughness exponent is close to one [6]. These results suggest that non-equilibrium statistical physics can play an important role providing insight into the problem of metal deformation.

As a result of the deformation process, 3D heterogeneous cell structures are formed, in which a nearly dislocation-free space (cell interior) is surrounded by high dislocation density walls (cell wall), as shown in Fig. 16.1b; see [8] for a review on dislocation patterning. A 2D fractal analysis of the cell structure shown in Fig. 16.1b reveals that the cell size distribution exhibits about one decade of power-law behavior [9], and a similar fractal structure has been found in a simulation of a 2D stochastic dislocation dynamics model [10]. The exponent characterizing the power-law decay in the size distribution is called the fractal dimension whose value is quite close to that found in a standard random percolation (SRP) problem. This power-law behavior indicates that the system might be in a scale-invariant critical state. Using a non-linear stochastic differential equation for the evolution of the total dislocation density, Hähner et al. [9,11] suggested that the formation of a fractal structure is associated with a noise-induced transition, where the noise intensity is a function of internal and external stresses, the average dislocation density, etc; see also [12,13].

Thomson and Levine [14] suggested that when a cell structure exists in the system, the slip line burst event is a percolation event due to the transport of mobile dislocations through the cell structure. Using a simple strain transport model called the strain percolation (SP) model, they demonstrated that a deforming metal can be considered as a self-organized critical system [14,15]. The SP model addresses how strain propagates through a cell structure and is relevant for strain propagation restricted to its initial slip plane. Subsequent study [16] showed that near the critical point, the geometrical aspects of the model (e.g., the mean number of strained cells, correlation length, etc) are well explained by the SRP theory [17]. This result also illustrates that a statistical approach could provide valuable insight into the problem of metal deformation. It should be noted that strain is an additional variable not present in the standard percolation problem.

In this paper, we present the strain-related critical behavior, discuss finite-size effects associated with long-transient behavior, and explore self-organized critical behavior in the system. In Sect. 16.2, we present the strain percolation model in detail and Sect. 16.3 contains our simulation results. Summary and conclusion are given in Sect. 16.4.

16.2 Model

First, we assume that a dislocation cell structure like the one shown in Fig. 16.1b exists in a metal single crystal and the system is in a stress-strain balanced, quiescent state, and then, as a response to an external stress increment, $\delta\sigma_{\mathrm{ext}}$, a strain, s_0, is nucleated at some weak cell on a slip plane. The nucleation event may occur several places in the structure, but for simplicity, we consider one nucleation event. The additional strain induced in a cell produces local stress fluctuations and drives the system into a state of strain. The strain fluctuations propagate through the cell structure by some strain transmission mechanisms, as shown schematically in Fig. 16.2, and eventually, the system evolves into a new strained state. For another external stress increment, the system is assumed to repeat this process.

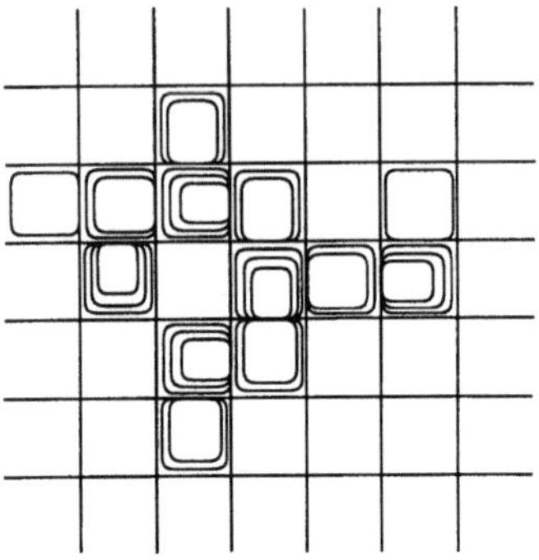

Fig. 16.2. A schematic view of strain propagation by a bowing-out mechanism starting from the initiating cell at the center. Square grids represent a cell wall structure

We consider two different mechanisms of strain propagation from a strained cell to unstrained neighboring cells. In the first case, a relatively stable wall can act as a source of new dislocations as well as a barrier, and the strain s propagates by activating sources in such walls. Since s is equivalent to a small number of incident dislocations expanding in the strained cell, the stress acting on the wall barrier between the two cells is proportional to the number of incident dislocations [1]. For this case, the magnitude of strain induced in an unstrained cell is assumed to be linearly proportional to the strain in a neighboring strained cell. The proportionality factor is a stochastic variable that reflects the distribution of sources within the walls. In the second case, as an additional mechanism, unstable locks within a wall may be broken by nearby incident dislocations leading to a large localized strain. We will consider here only the simpler first case.

The law of strain propagation from a strained cell to a neighboring unstrained cell is thus given as

$$s^* = sP\eta \,, \tag{16.1}$$

where s^* is the number of dislocations induced in the previously unstrained cell and s is the number in its strained neighbor. The percolation control parameter P represents, on the average, how efficiently the cell walls can transmit strain from a strained cell to an unstrained one. η is a random number with $0 \leq \eta \leq 1$, which reflects the highly random character of dislocation cell walls.

Recently, a physical basis for the somewhat intuitive strain transmission law given in (16.1) has been established by Thomson et al. [18]. Using random distributions of pinning points on the dislocations of the cell walls, they derived the percolation control parameter, P, as a function of physical variables such as length of dislocation (cell size), number of cell wall dislocations in front of a small dislocation pileup in a strained cell, external stress, etc. and showed the presence of a critical flow stress where P approaches its critical value.

In simulations, when an unstrained cell has more than one strained neighbor, the amount of strain transmitted to the unstrained cell, s^*, takes the largest value among the contributions from its neighbors. Due to work hardening that occurs in strained cells, additional straining of already strained cells becomes difficult, so that growth of a strained cluster is assumed to take place only on its periphery (i.e., in unstrained cells). Since dislocations are discrete entities, the minimum amount of strain that can be induced in an unstrained cell is unity; if $s^* < 1$ in a simulation, we set $s^* = 0$. In a real metal, the maximum amount of strain that can be induced in a cell, s_{m}, depends on the dislocation wall properties. For simplicity, however, we do not impose any restriction on s_{m}.

Our simulations have been carried out in a 2D square lattice, without cell size variation (see Fig. 16.2) since a simulation with a cell structure like one shown in Fig. 16.1b is a formidable task. The lateral system size is L, and all the length scales in the system are measured in units of a linear cell size $r_{\mathrm{c}} = 1$ which may be considered the mean cell size of a cell size distribution found in the sample. A single cluster is grown from the center of the system until growth stops. Since we consider slip on only a single slip plane, 2D is the relevant dimension for this problem. Although this growth algorithm is similar to that of Leath [19], it is a correlated percolation problem unlike other random percolation problems since the probability that a cell is strained depends on the strain of its neighbors as given in (16.1).

16.3 Results

16.3.1 Universal Scaling

Any accurate, statistical analysis of a critical system starts with determining the precise location of the critical point in the thermodynamic limit. In Fig. 16.3a, we show that in the limit of $L \to \infty$, the critical point is $P_{\mathrm{c}} = 1.306(2)$ for the initial strain $s_0 = 2.2$, where $P_{\mathrm{c}} - P_{\mathrm{c}}(L) \sim L^{-1/\nu}$ is used and $P_{\mathrm{c}}(L)$ is a size-dependent critical point that results in the maximum non-spanning cluster size for a given L. Other methods like a power-law fit, etc also result in the same critical point. It is also found that the value of P_{c} is independent of s_0 in the infinite-size limit.

Figure 16.3b shows a spanning cluster at the critical point, which is composed of strained cells, and the color scale represents the magnitude of strain developed in a cell. The general shape of the cluster is fractal with the fractal dimension $d_{\mathrm{f}} = 1.89(2)$ [16]. A large strain is developed in some regions – dark color regions, but decays rapidly as it propagates. It is found that the mean strain defined as the strain per strained cell is $\langle s \rangle_{\mathrm{c}} \approx 2$ at the critical point and the large area of grey in the figure indicates that the vast majority of cells have strain $s \approx 2$. For larger initial strain, $s_0 = 10$, the region

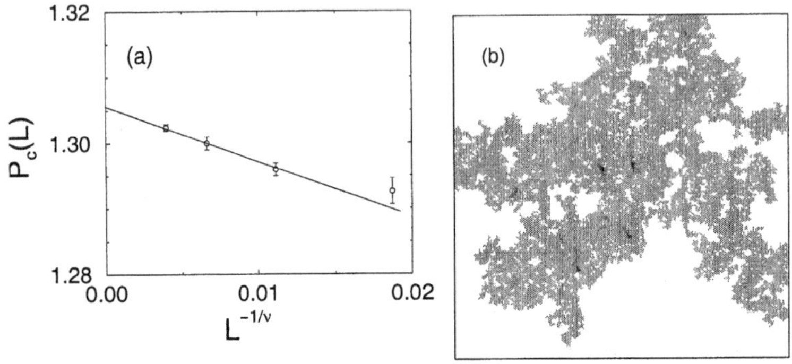

Fig. 16.3. (**a**) Extrapolation of the critical point to $L = \infty$ with the correlation length exponent, $\nu = 4/3$, yielding $P_c = 1.306(2)$. (**b**) Spanning cluster composed of strained cells at the critical point. Lighter colors represents smaller strain, while darker color represents larger strain in a cell. In both figures, $s_0 = 2.2$

near the origin, where a large strain is developed, decays very rapidly to the asymptotic value $\langle s \rangle \approx 2$.

Figure 16.4 shows critical power law behavior of the total strain defined as $T(P) = \langle \sum_{\boldsymbol{r}} s(\boldsymbol{r}, P) \rangle$. Near the critical point, the total strain for $s_0 = 2.2$ and 10 is well described by a power-law function,

$$T(P) \sim |P_c - P|^{-\chi} \tag{16.2}$$

with $\chi = 2.35(2)$. The saturation for small $P_c - P$ is due to finite-size effects. Note that in the previous study [16], it was found that the cluster size S exhibits universal behavior with $S(P) \sim |P_c - P|^{-\gamma}$ and $\gamma = 2.35(7)$; according to the SRP theory, $\gamma = 2.389$. The result that $\chi \approx \gamma$ implies that asymptotically the total strain also exhibits the same universal behavior as the cluster size, regardless of the value of s_0. But the size effects caused by a different initial strain become noticeable. For instance, as can be seen in Fig. 16.4, the scaling regime for $s_0 = 2.2$ is about $P_c - P \leq 0.1$, whereas for $s_0 = 10$, the regime becomes narrower ($P_c - P \leq 0.02$) and a larger system size is needed to observe such a power law.

It turns out that the initial-strain dependence appears as a correction to the universal scaling. That is, $T(P) \sim |P_c - P|^{-\gamma} (T_1 + a|P_c - P|^{\kappa})$, where $T_1 \approx 2.2$ and $\kappa (> 0)$ is a correction exponent with a constant a. It is found that $\kappa \approx 3/4$ and ≈ 4 for $s_0 = 2.2$ and 10, respectively [20]. Thus, (16.2) with $\chi = \gamma$ is asymptotically correct since the correction term can be safely ignored near the critical point. These results imply that for sufficiently large L, the total strain has a scaling relation,

$$T(P, L) \sim L^{\gamma/\nu} Q(L/\xi) \,, \tag{16.3}$$

where the scaling function is $Q(x) \sim x^{-\gamma/\nu}$ for $x \gg 1$ and $Q(x) \sim$ const. for $x \ll 1$ with the correlation length, $\xi(P) \sim |P_c - P|^{-\nu}$. Indeed, we obtained

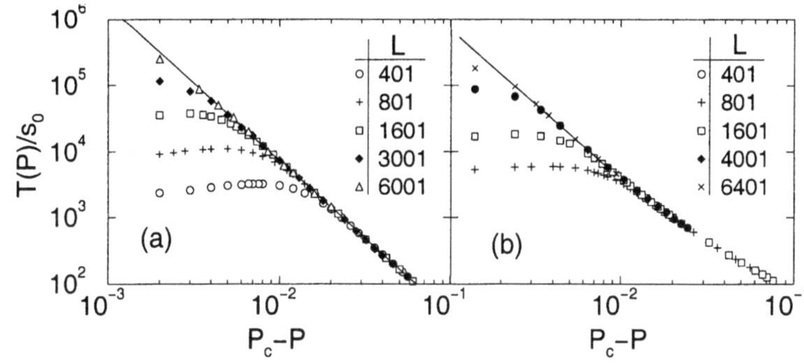

Fig. 16.4. Critical power-law behavior of total strain for (**a**) $s_0 = 2.2$ and (**b**) $s_0 = 10$ with $P_c = 1.306$. The slope of the solid lines is $-2.35(2)$

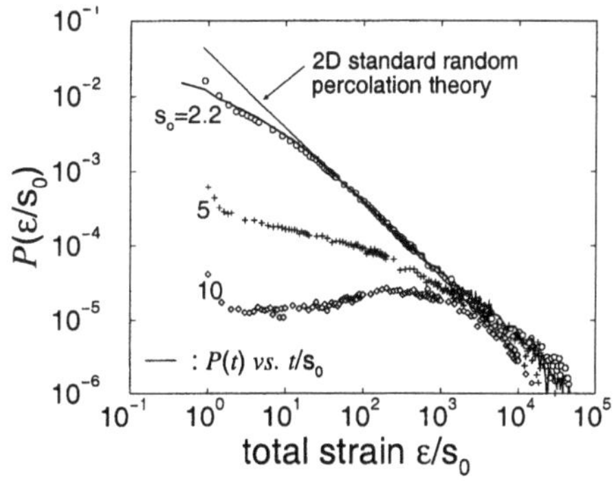

Fig. 16.5. Total strain distribution for $s_0 = 2.2, 5$, and 10 with $L = 801$ when $P_c - P = 0.0074$. The solid line is from 2D SRP theory with slope $= -1$. The curved line is for the cluster size distribution, $\mathcal{P}(t)$, of having size, t, for $s_0 = 2.2$; the cluster size t is also divided by s_0 for comparison

a nice scaling collapse using (16.3), and this result clearly demonstrates that the total strain displays asymptotically universal behavior regardless of the value of s_0.

The finite-size effects caused by the initial strain seem to appear severely in the total strain distribution, $\mathcal{P}(\varepsilon)$, and in the cluster size distribution, $\mathcal{P}(t)$, where the total strain $\varepsilon = \sum_{\boldsymbol{r}} s(\boldsymbol{r})$ and t is a cluster size. Figure 16.5 shows the total strain distribution $\mathcal{P}(\varepsilon)$ for three different values of s_0 with one cluster size distribution for $s_0 = 2.2$. It is found that when $s_0 = 2.2$, the size distribution exhibits a power-law decay with $\mathcal{P}(t) \sim t^{-(\tau-1)}$ and $\tau - 1 \approx 1$, which is well explained by the SRP theory ($\tau = 2.055$) [17]. This is also the case for the total strain distribution when $s_0 = 2.2$, i.e., $\mathcal{P}(\varepsilon) \sim \varepsilon^{-1}$.

On the other hand, when $s_0 > 2.2$, such power-law behavior does not appear until $\varepsilon/s_0 \sim 4 \times 10^3$. We observed a similar behavior in the cluster size distribution for such cases. Although the asymptotic scaling found in the total strain suggests that in the large-size limit, all the distributions display the same universal behavior regardless of s_0, it is not possible to confirm such behavior from simulations due to long-transient effects which become even stronger as s_0 increases.

According to the SRP theory, the density-density correlation function, $C_{\mathrm{g}}(r)$, also exhibits critical power-law behavior with $\beta/\nu = 5/48$;

$$C_{\mathrm{g}}(\boldsymbol{r}) = \langle \rho(0)\rho(\boldsymbol{r})\rangle \sim r^{-2\beta/\nu}\mathcal{G}(r/\xi)\,, \tag{16.4}$$

where the density $\rho(\boldsymbol{r}) = 1$ if a site $\boldsymbol{r}$ is occupied, otherwise it is 0. The scaling function given in (16.4) is $\mathcal{G}(r/\xi) \sim$ const. for $r \ll \xi$ and $\sim \exp[-r/\xi]$ for $r \gg \xi$. It is important to keep in mind that in a SRP problem, a site is occupied randomly with an equal probability and thus, there is no memory (a delta-correlation) between any two events of site occupation. On the other hand, in the SP problem, the probability that a cell is strained is not uniform and there is non-zero correlation between a newly strained cell and neighboring strained cells because of the strain transmission law given in (16.1). However, due to

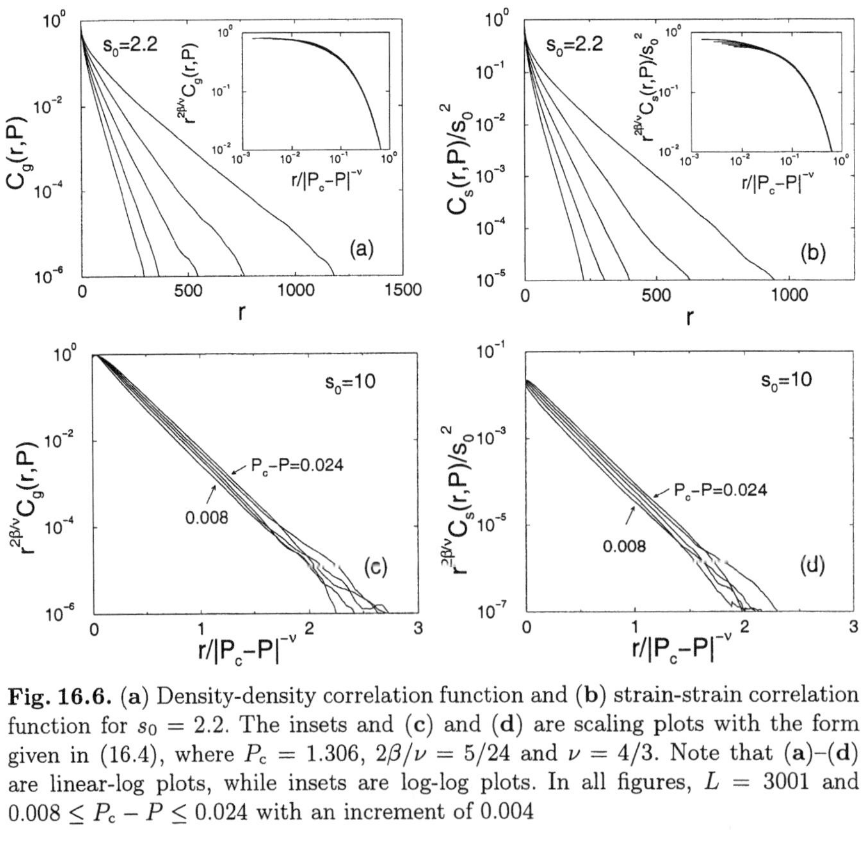

Fig. 16.6. (**a**) Density-density correlation function and (**b**) strain-strain correlation function for $s_0 = 2.2$. The insets and (**c**) and (**d**) are scaling plots with the form given in (16.4), where $P_c = 1.306$, $2\beta/\nu = 5/24$ and $\nu = 4/3$. Note that (**a**)–(**d**) are linear-log plots, while insets are log-log plots. In all figures, $L = 3001$ and $0.008 \le P_c - P \le 0.024$ with an increment of 0.004

the stochastic nature of (16.1), the memory effect caused by the transmission law may persist only up to some finite distance.

As can be seen in Fig. 16.6a, the density-density correlation function displays an exponential decay with $C_g(r) \sim \exp[-r/\xi]$. For small r, although we do not observe a clear, universal power-law decay with slope $= -2\beta/\nu$ even for $P_c - P = 0.008$, the scaling relation given in (16.4) for a SRP problem appears to be valid for this case; see the inset of Fig. 16.6a. In a similar way, a strain-strain correlation function is defined as $C_s(r) = \langle s(0)s(\boldsymbol{r})\rangle$. For large r, the strain correlation function exhibits the same exponential decay with $C_s(r) \sim \exp[-r/\xi]$; see Fig. 16.6b. It turns out that (16.4) is also a reasonable scaling form for $C_s(r)$, as shown in the inset of Fig. 16.6b, even though there seems to be a systematic deviation in the region of $r/|P_c - P|^{-\nu} \ll 1$ due to the short-range strain correlation resulted from the transmission law. For a larger strain, $s_0 = 10$, we observed the same exponential decay in the correlation functions as that for $s_0 = 2.2$, but for the range of $P_c - P$ and L that we used, scaling is relatively poor; see Fig. 16.6c and d. Nevertheless, the poor scaling may result from the long-transient behavior mentioned before and thus, we believe (16.4) is the asymptotic scaling form for the correlation functions although the data are not conclusive.

Figure 16.7a shows the strain decay function $\langle s(r)\rangle$ for $s_0 = 2.2$ and various values of $P_c - P$, where r is the distance from the initiation point to a strained cell. The strain decay function describes how the initial strain s_0 nucleated at the initiation point propagates through cells and decays. As the system approaches the critical point, a critical power-law decay becomes manifest for small r with $\langle s(r)\rangle \sim r^{-\alpha}$ and $\alpha \approx 0.28$, whereas for large r, $\langle s(r)\rangle \sim \exp[-r/\xi(P)]$. Thus, $\langle s(r)\rangle \sim r^{-\alpha}$ for $r \ll \xi$ and $\sim \exp[-r/\xi]$ for $r \gg \xi$.

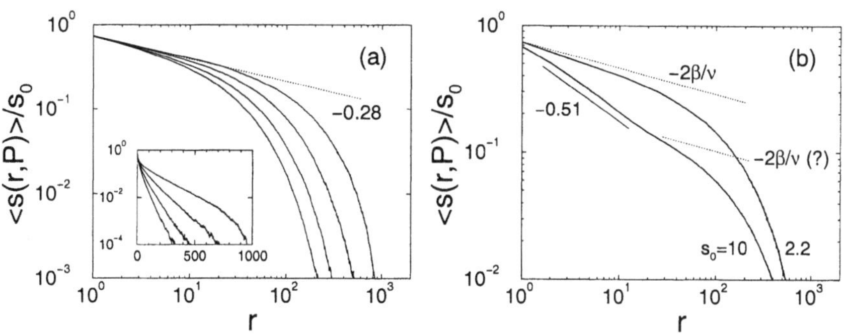

Fig. 16.7. (**a**) Strain decay function for $s_0 = 2.2$ and $0.004 \leq P_c - P \leq 0.016$ with an increment of 0.004 (from left to right). A power-law fit to the data for $P_c - P = 0.004$ yields a slope, $\alpha \approx 0.28$ and the inset is a linear-log plot. (**b**) Comparison between $s_0 = 2.2$ and 10 when $P_c - P = 0.004$, where $2\beta/\nu \approx 0.208$. In both figures, $L = 2001$

The exponent α seems to depend on s_0, as shown Fig. 16.7b. However, deeper insight into the small r region comes from considering the universal scaling found for the total strain. Since the total strain is the sum of all the strain developed in cells, in the critical regime,

$$
\begin{aligned}
T(P) &= \sum_{\boldsymbol{r}} \langle s(\boldsymbol{r}, P) \rangle \\
&\approx 2\pi s_0 a_1 \int_{r_c}^{\xi} r^{1-\alpha} dr + s_0 \sim \xi^{(2-\alpha)} \sim |P_c - P|^{-\nu(2-\alpha)}
\end{aligned}
\tag{16.5}
$$

with a constant a_1. Note that for $\alpha = 0.28$, the exponent $\nu(2-\alpha) \approx 2.3$ and is close to χ. This result implies that as $P \to P_c$ with sufficiently large L, α will approach the asymptotic value, $2\beta/\nu \approx 0.208$. This power law decay is the same as what we observed in the correlation functions, $C_g(r)$ and $C_s(r)$. Therefore, for $s_0 = 10$, a power law decay with slope $= -0.51$ looks like a transient before crossing over to a true asymptotic behavior denoted by a dotted line in Fig. 16.7b. It is not possible to observe such asymptotic behavior in computers available to us due to a huge computational demand.

The results of Figs. 16.4–16.7 illustrate that the system displays systematic, long-transient behavior which depends on the value of the initial strain assumed to be nucleated in some weak cell. Since the total number of cells used in the simulations is much larger than what we find in a deforming metal, size effects will be important in determining the stress-strain relation in a deforming metal. Thus, the system will exhibit various responses to the external stress, depending not only on the character and property of dislocation cell walls and but also just as importantly, on the number of cells formed in a cell structure.

At the critical point, the total strain displays a power-law size dependence, $T(L) \sim L^{d_f}$ like the cluster size ($S(L) \sim L^{d_f}$), with the fractal dimension $d_f = 1.896(3)$; see the inset of Fig. 16.8. This value is consistent with the SRP result ($d_f = 1.8958$), and it appears that d_f does not depend on the initial strain s_0. It is also found that at the critical point, $T(L)/S(L) \approx 2.2$ for sufficiently large system and the ratio seems not to depend on the initial strain. A fractal is a geometrical object whose dimension is smaller than the embedding dimension of the system (in this case, the embedding dimension is $d_E = 2$). Since the total strain is not a geometrical object, it is not trivial to understand the physical origin of $d_f < d_E$. It turns out that the fractal behavior in the total strain is the direct outcome of a self-organizing character of the strained cells, which will be discussed in the next section.

Since $T(P_c, L)/L^2 \to 0$ as $L \to \infty$, in order to have a finite non-zero strain per cell in the limit, the percolation control parameter P should be $P > P_c$; in this region, the fractal dimension of T becomes $d_f = d_E$. And our finding of $T(P_c, L) \sim L^{d_f}$ also implies that the total strain behaves as

$$
T(P, L) \sim L^{d_f} \mathcal{F}(L/\xi) \tag{16.6}
$$

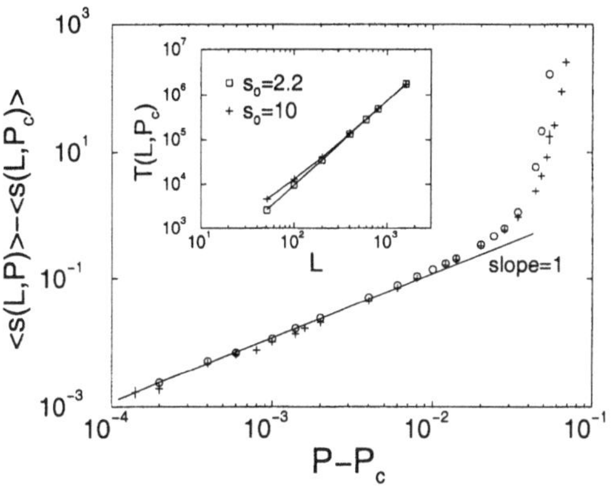

Fig. 16.8. Linear increase of the mean strain from its critical value near P_c for $s_0 = 2.2$ and 10 and $L = 801$. The inset shows fractal behavior of total strain at the critical point; a power-law fit to the data for $s_0 = 2.2$ yields a slope, $d_f = 1.896(3)$

for small $|P - P_c|$. The scaling function $\mathcal{F}(x)$ is constant for $x \ll 1$ and $\mathcal{F}(x) \sim x^{2-d_f}$ for $x \gg 1$. Thus, for an infinite size of system, the strain per cell is $T(P, L)/L^2 \sim (P - P_c)^\beta$ with $\beta = \nu(2 - d_f) = 5/36$, i.e., the SP theory predicts that strain per cell is well behaved at P_c, not singular and in the thermodynamic limit, it increases with a power law as P increases. Experimental observations of surface slip line structure also strongly suggest that the relevant physical state is just above P_c, so this well behaved character is crucial to our understanding of the physical system.

Another interesting feature in Fig. 16.8 is that the mean strain defined as strain per strained cell increases linearly from its critical value for a relatively narrow range of $P - P_c$ and then diverges quickly. This divergence (or avalanche) in the mean strain may not be physical in a real system. That is, the unphysical feature can be removed by setting $s_{\max} = finite$. Since a large deforming system demands a finite strain from a percolation event, this can only be accomplished by the physical system inhabiting the region between the critical and avalanche points. So this narrow region is very important, physically. Further, depending on the cutoff value of $s_{\max}$ the avalanche rise may or may not be a real physical effect.

16.3.2 Self-Organized Criticality

As mentioned before, at the critical point, the strain propagation stabilizes to a small but finite value that is close to the minimum stable value of strain. Since the minimum amount of strain that can be induced in a cell is $s^* = 1$, for any trial transmission, the value of the strain in a strained cell must be such that from (16.1),

$$P_c \langle \eta \rangle \langle s \rangle \geq 1 . \tag{16.7}$$

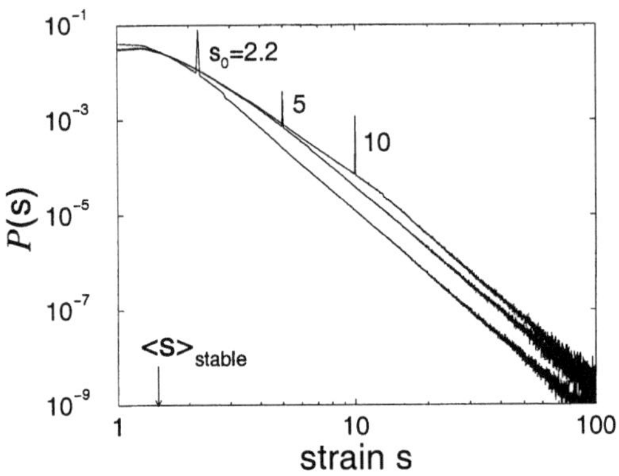

Fig. 16.9. Probability distribution of having strain s at the critical point for $s_0 = 2.2$, 5, and 10 with $L = 801$. For $s > s_0$, the distribution exhibits a power-law decay, $\mathcal{P}(s) = \mathcal{P}_0\, s^{-4.4}$ with a constant $\mathcal{P}_0$. The sharp peaks are due to the initial strain s_0, and the arrow is at $\langle s\rangle_{\mathrm{stable}} = 1.5$

Thus, the minimum stable value of strain is given approximately by $\langle s\rangle_{\mathrm{stable}} \approx 2/P_\mathrm{c} \approx 1.5$, where we assume that any possible correlation between P_c and s is negligibly small at the critical point. This assumption seems to be reasonable because the values of P_c and $\langle s\rangle_\mathrm{c}$ have virtually no dependence on the initial strain s_0. The result of $T(L)/S(L) \approx 2.2$ at the critical point explains why for the initial strain $s_0 = 2.2$, the system shows relatively clear universal behavior compared to other cases having larger initial strain.

Figure 16.9 shows the strain probability distribution function, $\mathcal{P}(s)$ at the critical point, where $\mathcal{P}(s)\,ds$ is the average number of cells having the strain, s. As can be seen in Fig. 16.9, $\mathcal{P}(s)$ varies slowly near the minimum strain. For larger strains ($s > s_0$), however, the strain distribution exhibits a fast power-law decay with $\mathcal{P}(s) \sim s^{-4.4}$. As predicted above, $\langle s\rangle_{\mathrm{stable}}$ is very close to 1.5. A strain $s > \langle s\rangle_{\mathrm{stable}}$ is most likely to decay to a lower value when transmitted to an unstrained cell. At the critical point, the number of cells having strain larger than some value $\tilde{s}$ are $S(L)\int_{\tilde{s}}^{\infty}\mathcal{P}(s)\,ds = n(\tilde{s})S(L)$, where the cluster size $S(L) \sim L^{d_\mathrm{f}}$ and the number density $n(\tilde{s}) \approx n_0\,\tilde{s}^{-3.4}$ for $\tilde{s} > s_0$. When $s_0 = 2.2$, $\mathcal{P}_0 \approx 0.33$ and $n_0 \approx 0.097$, and thus, the number density $n(\tilde{s})$ is slightly smaller than 0.025 for $\tilde{s} = \langle s\rangle_{\mathrm{stable}}$ and $n(\tilde{s}) \approx 0.0067$ for $\tilde{s} = s_0$. We therefore predict that experimental measurements of slip line heights will follow this same distribution. Work on measuring this distribution for incrementally deformed single-crystal aluminum is in progress and current results show that the predicted magnitudes are consistent with the predictions [21].

Therefore, even though a few cells develop large strain, most of the cells are near the minimum or at least marginally stable state regardless of the initial conditions, a behavior similar to that found in other self-organized critical systems [15]. The universal behavior found in the system is the consequence of self-organizing critical behavior of cells because the variation of strain in a huge population of the SOC cells is quite small. For larger initial strains ($s_0 > 2.2$), the relatively extended behavior of $\mathcal{P}(s)$ between $\langle s \rangle_{\text{stable}}$ and s_0 seems to cause a long transient effect before crossing over to a true asymptotic behavior.

16.4 Summary and Conclusion

Using a strain percolation model, which is relevant for a deformation stage where a dislocation cell structure is formed, we examine strain behavior of the model as a function of two key model parameters, the initial strain and percolation control parameter. The former is the amount of strain nucleated in some weak cell and the latter reflects a cell wall response to nearby incident dislocations, and both are assumed to be a function of the macroscopic stress and strain. Our key physical picture for the strain transmission from one cell to another is summarized as a stochastic linear response of dislocations in one cell wall to other dislocations generated in a neighboring cell.

The criticality of our model is characterized by a critical point and some critical exponents. It is found that the critical point has virtually no dependence on the initial strain, implying that the system evolves into an initial strain-independent, critical state. Although the system exhibits long-transient behavior depending on the level of the initial strain (which implies systematic finite-size effects), as the percolation control parameter increases, the overall system progresses from a subcritical regime of essentially negligible macroscopic strain to a critical state above which the strain is non zero in the thermodynamic limit, and finally to strain avalanche.

As the system approaches the critical point, the majority of the strained cells in a strained cluster take on a strain value which is independent of the initiating strain. This strain value is that which is marginally capable of being propagated indefinitely through the system, and is only slightly larger than the strain below which an initiating strain cannot propagate at all. One of our central findings is that near the critical point, the total strain exhibits critical power-law behavior with universal critical exponents, which are related to standard percolation theory. This connection to standard percolation theory exists even though that theory contains no variable corresponding to the strain variable of our model. This universal behavior seems to result from a huge number of self-organized critical cells having a marginably stable value of strain.

Although we have not made the detailed connection between our model and actual metal deformation in this paper, the strain percolation model is intended to correspond to the fine slip which appears at all stages of deformation.

References

1. D. Hull and D.J. Bacon: *Introduction to dislocations* (Butterworth-Heinemann, Woburn 1999)
2. M.-C. Miguel, A. Vespignani, S. Zapperi, J. Weiss, and J.-R. Grasso: Nature **410**, 667 (2001)
3. J. Weiss, J.-R. Grasso, M.-C. Miguel, A. Vespignani, and S. Zapperi: Mater. Sci. Eng. A **309–310**, 360 (2001)
4. M.-C. Miguel, A. Vespignani, S. Zapperi, J. Weiss, and J.-R. Grasso: Mater. Sci. Eng. A **309–310**, 324 (2001)
5. J.G. Sevillano, I.O. Arizcorreta, and L.P. Kubin: Mater. Sci. Eng. A **309–310**, 393 (2001)
6. S. Zapperi and M. Zaiser: Mater. Sci. Eng. A **309–310**, 348 (2001)
7. H. Mughrabi, T. Ungar, W. Kienle, and M. Wilkens: Philos. Mag. A **53**, 793 (1986)
8. L.P. Kubin, in *Materials science and technology: a comprehensive treatment.* Ed. by R.W. Cahn, P. Haasen, E.J. Kramer (VCH, Weinheim 1993) Vol. 6, Chap. 4
9. P. Hähner, K. Bay, and M. Zaiser: Phys. Rev. Lett. **81**, 2470 (1998)
10. I. Groma and B. Bako: Phys. Rev. Lett. **84**, 1487 (2000)
11. M. Zaiser and P. Hähner: Mater. Sci. Eng. A **270**, 299 (1999); M. Zaiser, K. Bay, and P. Hähner: Acta Mater. **47**, 2463 (1999)
12. P. Hähner: Acta Matter. **44**, 2345 (1996); Appl. Phys. A **62**, 473 (1996)
13. M. Zaiser: Mater. Sci. Eng. A **309–310**, 304 (2001)
14. R. Thomson and L.E. Levine: Phys. Rev. Lett. **81**, 3884 (1998)
15. P. Bak, C. Tang, and K. Wiesenfeld: Phys. Rev. Lett. **59**, 381 (1987)
16. R. Thomson, L.E. Levine, and D. Stauffer: Physica A **283**, 307 (2000)
17. D. Stauffer and A. Aharony: *Introduction to Percolation Theory.* (Taylor and Francis, London 1994)
18. R. Thomson, L.E. Levine, and Y. Shim, to be published.
19. P.L. Leath: Phys. Rev. B **14**, 5046 (1976)
20. Y. Shim, L.E. Levine, and R. Thomson: Mater. Sci. Eng. A **309–310**, 340 (2001)
21. L.E. Levine, R. Thomson, M.F. Savage, D.E. Kramer, and Y. Shim: "Single-Crystal Plasticity: Statistical Physics and Experiments", in *Plasticity, Damage and Fracture at Macro, Micro and Nano Scales.* Ed. by A.S. Khan and O. Lopez-Pamies (Neat Press, Fulton 2002)

17 Calculation of Static and Thermal Displacements in Metallic Alloys

Y. Puzyrev and J.S. Faulkner

Alloy Research Center, Department of Physics, Florida Atlantic University, Boca Raton, FL 33431, USA

Abstract. Molecular Dynamics calculations of atomic displacements were performed to study size effects in binary metallic alloys. The copper-gold alloy was chosen as the system of interest due to a large size mismatch and the existence of well developed embedded-atom potentials. Static displacements are shown to depend significantly on temperature while thermal displacement are shown to be equal for both components of alloy, which was an unjustified assumption made by X-ray diffraction experimentalists.

17.1 Theory

The explanation of the properties of solids based of density functional theory local density approximation (DFT-LDA) [1] calculations of the electronic states has been so successfull that, on the occasion that there is not complete agreement with experimental results, scientists look for something that was left out of the calculation. Problems with understanding the photoemission spectra of copper palladium alloys led to the suggestion that the difficulty with the DFT-LDA was due to ignoring the displacements of the atoms from their ideal lattice positions [2,3]. Most of the groups that are involved in developing first-principle theories for metallic alloys are now making an effort to incorporate atomic displacements (AD) into their methods [4,5].

17.1.1 Atomic Displacements

Calculations of AD have been done on large supercells [6] using embedded-atom potentials [7], and on small clusters of atoms with DFT-LDA calcualtions [8]. All of these calculations use a model that is colder than absolute zero in the sense that atoms have no motion other than their static displacements. They rely on the results of a small number of a EXAFS experiments to check their theories. Two of the most frequently quoted EXAFS papers which purport to measure the AD in *concentrated* metallic alloys reach conflicting conclusions about their general nature [8,9]. It would be extremely useful to have reliable evidence about the AD in metallic alloys, and the most accurate way to do this experimentally is to analyze the diffuse scattering data taken with X-rays at a synchrotron source such as the APS in Argonne, Illinois [10].

17.1.2 Atomic Size Effects

Atomic size effects have been studied by X-ray and neutron diffractionists for decades. The reduction in the intensity of the Bragg peak located at the scattering angle θ is given by the factor $\exp[\frac{-2(B+B')\sin^2\theta}{\lambda^2}]$, where B is the standard Debye–Waller factor and B' is a static Debye–Waller factor. These factors are proportional to mean square displacements of the atoms about their ideal sites due to the temperature-dependent excitation of phonons $< u_{th}^2 >$, and the mean square of the static atomic displacements $< u_{st}^2 >$. Since B' is independent of temperature, B and B' can be found individually from powder diffraction patterns measured at several temperatures. For the case of an alloy with very large size mismatch, copper-gold, it was found experimentally [11] that B' is equal to B at liquid nitrogen temperatures and $B' = B/3$ at room temperature. If copper and gold retained the sizes that they have in pure metals, B' would be twenty times as large as one observed. Of course, the B' only gives a measure of the mean square displacements, whereas the diffuse X-ray scattering measurements give much more detailed information on the AD. It appears obvious that calculations of AD that ignore all motions of the atoms other than static one are doomed to failure. Therefore we are calculation the displacements of atoms in an alloy at room temperature and above using molecular dynamics. From this we will obtain both static and thermal displacements. The arrangements of atoms that arise in the calculation will be used to calculate the diffuse scattering maps directly. We are using EAM potentials [7] and ParaDyn molecular Dynamics code [12].

17.2 Results of Calculations

Calculations were performed on 4000 atoms using a NPT ensemble with minimal image convention for force calculations. Hoover damping [13] was chosen as a temperature control mode. For each temperature and concentration number of MD steps must be allowed for thermal expansion until equilibrium is reached.

17.2.1 Thermal Expansion and Vegard's Law

From Fig. 17.1 we can see the difference in atomic sizes of copper and gold as well as the thermal expansion of these two elements. It is clear that MD predicts thermal expansions well, and therefore it can be used for the evaluation of thermal and static isplacements.

The dependence of the lattice constatnts on concentration is in good agreement with experiment as shown in (Fig. 17.2). This indicates that the EAM potentials describe the atomic sizes adequately. Note that the calculations are performed at the same temperature as the experiments, i.e. room temperature.

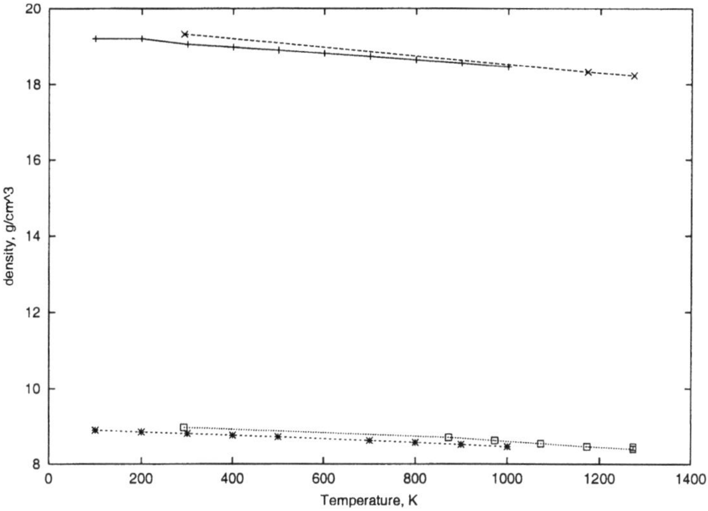

Fig. 17.1. Thermal expansion of copper and gold. Solid lines represent MD results. Dash lines show experimental data [14]. Gold has twice as high density as copper

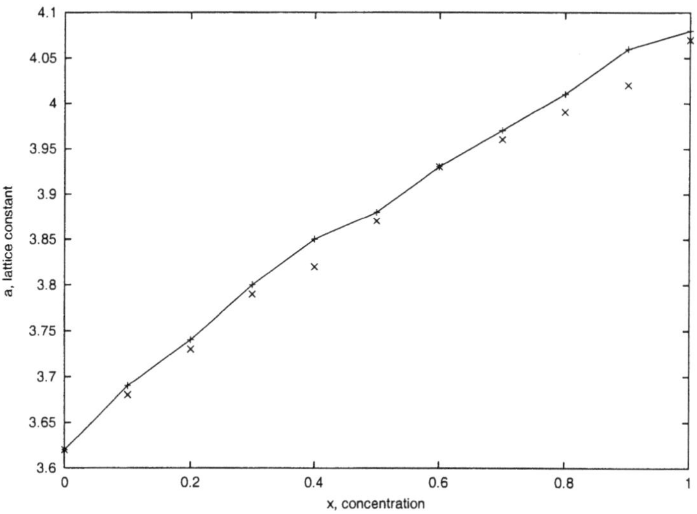

Fig. 17.2. Lattice parameter of $Au_{50}Cu_{50}$ alloy as function of concentration. Lines is our Molecular Dynamics result, Dots are experimental data [15,16]

17.2.2 Mean Square Displacements

As shown in Fig. 17.3 static displacement depend on temperature and thermal displacements are the same for both copper and gold. The reason for static displacements being so large is that our evaluation of them is not consistent

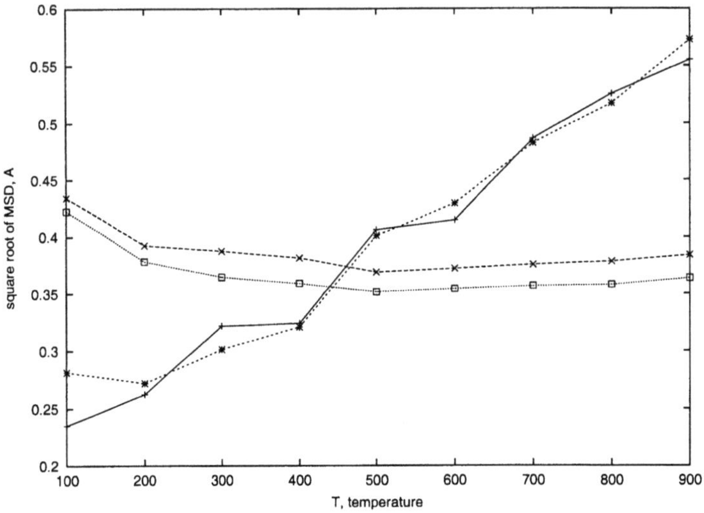

Fig. 17.3. Thermal and static displacements in $Au_{50}Cu_{50}$ alloy. Static displacements: –x– Gold; –o– Copper. Thermal displacements: -+- Copper; –*– Gold

with the conventional experimental definition. In the case of molecular dynamics, the atomic configuration is available for each simulation step. Hence it is intuitive to average the positions of each particular atom over time and find the mean square displacement as a distance between this average position and the corresponding site position in the ideal lattice.

17.3 Conclusions

- Simple classical Molecular Dynamics with embedded-atom potentials predicts well thermal expansion and shows behavior consistent with Vegard's law.
- Thermal displacements are the same for both copper and gold, which justifies assumption that experimentalists make.
- Static displacements are indeed strongly dependent on temperature.

In the future we plan to:

- Reanalyze the calculation in a manner consistent with definitions used by diffractionists.
- Use the instanteneous atomic positions from MD calculation will be used to predict the diffuse scattering maps measured experimentally.
- Treat other alloys.

References

1. P. Hohenberg and W. Kohn: Phys. Rev. **136**, B864(1964); W. Kohn and L.J. Sham: Phys. Rev. **140**, A1133 (1965)
2. D. van der Marel, J.A. Jullianus, and G.A. Sawatzky: Phys. Rev. B **32** 6331 (1985)
3. P. Weightman, H. Wright, S.D. Waddington, D. van der Marel, G.A. Sawatzky, G.P. Diakum, and D. Norman: Phys. Rev. B **36**, 9098 (1987)
4. Z.W. Lu, S.H. Wei, and A. Zunger: Phys. Rev. B **44**, 3387 (1991)
5. P.E.A. Turchi, D. Mayou, and J.P. Julien: Phys. Rev. B **56**, 1726 (1997)
6. N. Mousseau and M.F. Thorpe: Phys. Rev. B **45**, 2015 (1992)
7. M.S. Daw, S. Foiles, and M.I. Baskes: Mater. Sci. Rep. **9**, 251 (1993)
8. G. Renaud, N. Motta, F. Lancon, and M. Belakhovsky: Phys. Rev. B **38**, 5944 (1988)
9. A.I. Frenkel, E.A. Stern, A. Rubstein, A. Voronel, and Y. Rosenberg: J. Phys. IV France **7**, C2–1005 (1997)
10. G.E. Ice et al.: J. of Phase Equilibria **19**, 529 (1998)
11. F.H. Herbstein, B.S. Borie, Jr., and B.L. Averbach: Acta Cryst. **9**, 466 (1956)
12. S.J. Plimpton and B.A. Hendrickson, in *Materials Theory and Modelling.* Ed. by J. Broughton, P. Bristowe, and J. Newsam, MRS Proceedings 291 (Pittsburgh, PA 1993) p. 37
13. W.G. Hoover: Phys. Rev. A **31**, 1695 (1985)
14. A. Buch: *Pure metals properties: a scientific-technical handbook* (Materials Park, Ohio: ASM International 1999)
15. J.O. Linde: Ann. Phys. (Leipzig) [Folge 5] **15**, 249 (1932); F.C. Nix and D. Macnair: Phys. Rev. **60** 597 (1941); J.B. Newkirk: J. Metals **5**, 823 (1953)
16. F. Lihl, H. Ebel, and G. Baumgartner: Z. Metallkd. **62**, 42 (1971)

18 Monte Carlo Simulation of the Surface Structure of Ge on Si(001)

L. Nurminen[1], F. Tavazza[2], D.P. Landau[1,2], A. Kuronen[1], and K. Kaski[1]

[1] Helsinki University of Technology, Laboratory of Computational Engineering, P.O.Box 9203, FIN-02015 HUT, Finland
[2] The University of Georgia, Center for Simulational Physics, Athens, GA 30602-2451, USA

Abstract. Large-scale Monte Carlo simulations are used to study the structural properties of small germanium islands on a silicon (001) surface. The interactions between particles are modeled using empirical potentials of the Stillinger–Weber and Tersoff forms. The suitability of the potentials for finite-temperature simulations of the Si(001) surface is examined and shortcomings are identified. The Ge islands on Si(001) show significant edge effects caused partly by the underlying substrate and to a lesser extent by lattice-mismatch induced strain in the islands.

18.1 Introduction

Mixed semiconductor systems have attracted great interest due to their important application possibilities in the microelectronic industry. The properties of germanium differ only slightly from those of silicon, which makes the Si-Ge system an ideal candidate for the study of mixed systems. The interesting phenomena related to surface reconstructions and strain relaxation involve length scales far beyond the reach of ab initio methods. Valuable insight into these otherwise intractable problems can be gained by using classical interatomic potentials in connection with large-scale simulations. Equilibrating semiconductor systems with complicated energy landscapes requires highly efficient simulation techniques. We find the Monte Carlo (MC) method to be the most suitable tool for investigating the structure of mixed Si-Ge systems.

Large-scale constant-pressure Monte Carlo simulations are employed to study the atomic scale structure of small Ge islands on a Si(001) substrate. The islands are observed to display significant edge effects indicating possible splitting of the islands beyond certain critical size. The splitting is predicted to occur in the direction perpendicular to the Ge dimer bonds. Dimer formation and relaxation of the surface are also discussed from the computational point of view.

18.2 Simulation Method

The Ge/Si(001) system is modeled using a slab containing 16 or 20 layers with the surface size varying from 8×8 to 64×64 atoms. The simulation unit cell is constructed along the [110], [1$\bar{1}$0] and [001] directions. Periodic boundary conditions are used in the x and y directions. In the simulations, the surfaces are initially given a 2×1 reconstruction or left bulk terminated. The structural energy of the system, E, is calculated from the atomic coordinates using the Stillinger–Weber (SW) potential [1] or one of the Tersoff potentials (T2 or T3) [2,3]. T2 and T3 are different parametrizations of the same functional form. Ge parametrization exists for the SW [4] and the T3 [5] models. At each MC step, we choose a particle at position $\boldsymbol{r}_i$ and attempt to displace it by a small amount to position $\boldsymbol{r}_i'$. The acceptance probability is given by the standard Metropolis form with the energy difference calculated using the structural energy E. After each particle in the system has been attempted, we randomly choose new linear sizes for the system (L_x', L_y', L_z'), slightly altered from the previous ones. The positions of the atoms are scaled by the relative change in the linear system size: $x' = xL_x'/L_x$, $y' = yL_y'/L_y$ and $z' = zL_z'/L_z$. The acceptance probability is again calculated using the Metropolis form, now using an effective Hamiltonian given by $H_{\text{eff}} = E - Nk_{\text{B}}T\ln(L_xL_yL_z)$ [4].

18.3 Comparison of Empirical Potentials

Originally, most of the empirical potentials designed for silicon were fitted to bulk properties leaving their applicability to surface studies questionable. Some comparative studies have been performed to test the strengths and limitations of the potentials (see e.g. [7]); however, most previous studies have concentrated exclusively on static or zero-temperature calculations. The motivation in our study is to examine the suitability of three widely used potentials, SW, T2 and T3, for finite-temperature simulations of the Si(001) surface.

Experimentally the Si(001) surface is observed to form a 2×1 reconstruction in which the surface atoms organize into parallel dimer rows [6]. For all three potentials, the formation of dimers is energetically favorable but in simulations the surface reconstructs to form only small domains of straight dimer rows separated by antiphase boundaries (dimer rows are shifted by one lattice constant relative to each other). The energy difference between a configuration with two antiphase domains (both 2×1 reconstructed) and an ideal 2×1 reconstruction is extremely small. Our simulations at $T = 0.05\,\text{eV}$ give the following values for *the energy difference per dimer row*: 0.20 eV for T2, 0.083 eV for SW, and −0.013 eV for T3. Thus the Tersoff potential T3 is found to favor the antiphase configuration over the experimentally observed 2×1 structure, which limits the suitability of this potential for simulation studies of the Si(001) surface.

In addition, the finite thickness of the simulation slab was found to influence the bulk properties of the system. The effects of the surface have generally been assumed to extend only a few atomic layers into the bulk, thus a 20-layer slab should be sufficiently thick to prevent the two surfaces

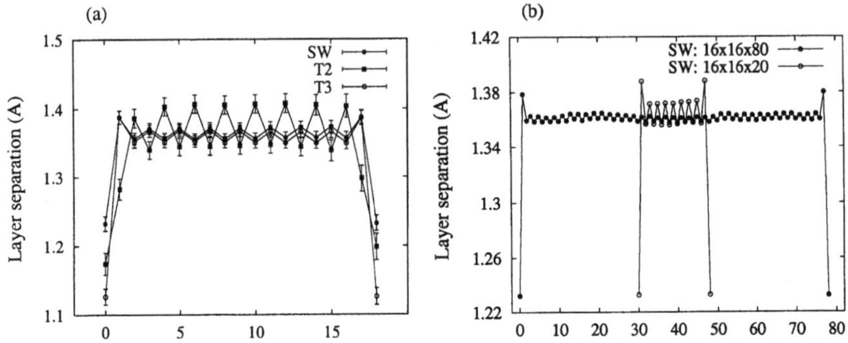

Fig. 18.1. Separation between atomic layers in the direction perpendicular to the (001) plane. (**a**) Comparison of different potentials. The system size is $16 \times 16 \times 20$ atoms. (**b**) Effect of slab thickness. The smaller slab has a thickness of 20 atomic layers and the larger slab 80 atomic layers. Temperature is 0.05 eV in both figures

from interacting with each other. Most measured quantities (average number of bonds, average bond length and bond angle, etc.) do reach a constant value starting approximately from the fifth substrate layer below the surface. However, we found that the average separation between atomic layers in the direction perpendicular to the (001) plane oscillates with a constant amplitude throughout the bulk. As can be seen from Fig. 18.1a, all three potentials produce similar oscillations, with SW and T3 giving identical results within statistical errors. Figure 18.1b shows that increasing the slab thickness reduces the amplitude of the oscillations. Such oscillations are not observed if periodic boundaries are imposed in all three directions, thus the oscillations are caused by the finite thickness of the slab. Surprisingly the amplitude does not decrease when moving toward the center of the slab, as would physically be expected of a surface effect. It should be noted that under the same conditions, T2 produces larger oscillations than T3 and SW. We know that the angular function of T2 is more flexible than those of T3 and SW [7]. This gives reason to believe that the inward relaxation of the first surface layer (which is in agreement with ab initio calculations [7]) produces a driving force for the oscillations that competes with the energy cost of distorting the bond angles and bond lengths. Thus careful attention should be paid to the selection of the potential and to its limitations when interpreting the results.

18.4 Ge Islands on Si(001)

As a physical application, we used the SW potential to study the relaxation of small Ge islands on a Si(001) surface. From experimental observations it is expected that islands larger than a certain critical size break apart due to growing strain in the island. Figure 18.2 shows the separation between

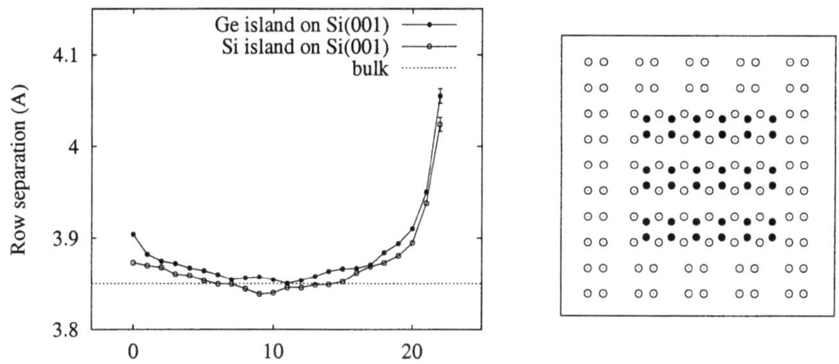

Fig. 18.2. Separation between dimer rows in the direction along the dimer bond for a Ge island (strained) and a Si island (unstrained) on Si(001). The size of the islands is 24 × 24 and the substrate size is 32 × 32. The atomic configuration of the island and first substrate layer is shown schematically on the right. The filled circles denote the top layer atoms (island) and the open circles atoms in the first substrate layer

vertical dimer rows (in the direction parallel to the dimer bonds) for a Ge island on Si(001) and a Si island on Si(001) at $T = 0.05\,\text{eV}$. In both cases the island is square shaped with a size of 24 × 24 atoms. The Si substrate underneath has a surface size of 32 × 32 atoms. The atomic configuration of the island and first substrate layer is shown schematically (for a much smaller system) on the right in Fig. 18.2. The asymmetric form of the curves is an effect caused by the substrate dimers. When a substrate dimer row is directly bonded to an edge of the island (right edge of the island in Fig. 18.2), the substrate atoms "pull" the edge rows away from the island, which causes the island to spread in the direction of the substrate dimer bonds. The other edge of the island is not directly attached to a substrate dimer row, thus we do not observe such a pulling effect. At this end, however, the difference between the Ge island and the Si island becomes visible. The row separation in the Si island stays very close to the bulk value of 3.85 Å, while the Ge island spreads due to strain.

18.5 Conclusion

We have performed large-scale Monte Carlo simulations to study the surface structure of Ge on Si(001). Empirical interatomic potentials are used in the simulations because the phenomena we are interested in requires large system sizes reaching up to 10^5 particles. The suitability of the potentials for use in finite-temperature simulations of the Si(001) surface was investigated. The results show that some widely used potentials suffer from problems that can influence the outcome of the simulations significantly.

The SW potential was used to study the relaxation of small Ge islands on Si(001). The islands were found to display notable edge effects due to two different reasons. If one edge of the island is directly bonded to a substrate dimer row, the island expands due to pulling from the substrate dimers. Lattice-mismatch induced strain in the island also causes spreading in the same direction, but this effect is observed to be much weaker. However, these results give indication that the splitting up of very large islands occurs in the direction perpendicular to the Ge dimer bonds since even small islands show spreading in this direction. In on-going work we are developing more efficient simulation techniques to be able to address these issues in more detail.

Acknowledgments

This work was supported in part by the Academy of Finland, project on Computational Research of Semiconductor Materials, Project No. 1169043 (Finnish Center of Excellence Program 2000–2005), and in part by NSF, Grant No. DMR-0094422. L.N. would also like to acknowledge support of the Finnish Cultural Foundation.

References

1. F.H. Stillinger, T.A. Weber: Phys. Rev. B **31**, 5262 (1985)
2. J. Tersoff: Phys. Rev. B **37**, 6991 (1988)
3. J. Tersoff: Phys. Rev. B **38**, 9902 (1988)
4. M. Laradji, D.P. Landau: Phys. Rev. B **51**, 4894 (1995)
5. J. Tersoff: Phys. Rev. B **39**, 5566 (1989)
6. Z. Zhang, F. Wu, M.G. Lagally: Annu. Rev. Mater. Sci. **27**, 525 (1997)
7. H. Balamane, T. Halicioglu, W.A. Tiller: Phys. Rev. B **46**, 2250 (1992)

Part IV

Liquids and Colloids

19 Simulating Semiconductor Liquids with Ab Initio Pseudopotentials and Quantum Forces

J.R. Chelikowsky, M. Jain, and J.J. Derby

Department of Chemical Engineering and Materials Science, Minnesota Supercomputing Institute, University of Minnesota, Minneapolis, MN 55455, USA

Abstract. One of the most difficult problems in materials science is simulating the microscopic liquid state. Often interatomic forces crafted from *classical potentials* are employed in such simulations. These potentials are fit to experimental data and may not replicate the true forces in the melt. Here we illustrate how *quantum* forces can be used to simulate liquids. These forces are determined within the pseudopotential-density functional method. This method is highly accurate and well tested for semiconductor in the solid state, but has only recently been applied to liquids. We will illustrate this approach for Si, GaAs and ZnTe liquids. For these liquids, we will present results for the microstructure, the diffusion constants and the electronic properties.

19.1 Introduction

Developing an understanding of the liquid state (l-state) is a formidable challenge, especially for liquid semiconductors. In general, semiconductors have been extensively studied in the crystalline phase, but the liquid properties for these semiconductors have not received similar attention. Although liquid semiconductors are technologically important (e.g. zone refining in laser annealed surfaces or crystal growth), the high melting points and the high reactivity of these liquids make experimental examinations difficult [1].

Theoretical studies of liquid semiconductors have also been somewhat limited owing to the complexity of the problem, e.g., large numbers of atoms with little symmetry [2,3]. Given the dynamical nature of the liquid state, one needs to construct representative ensembles [4]. The ensemble must be sufficiently large to capture the essential features of the l-state, but not so large as to preclude an accurate numerical simulation and it should reflect a user specified temperature and pressure.

Obtaining accurate interatomic interactions is the most difficult aspect in the process of simulating an accurate ensemble. One would like to treat the interactions in a fully quantum mechanical fashion. In this way, interatomic interactions involving metallic and covalent bonding, hybridization changes, or charge transfers can be handled correctly.

Pseudopotentials can be used to calculate quantum forces within the liquid state. This approach allows a number of fundamental issues to be addressed without resort to empirical approaches. Empirical approaches are particularly difficult for liquid semiconductors. Semiconductors like silicon are known to have strong "bond bending" forces. These forces correspond to three body interactions and cannot be represented by pairwise interactions [5]. Many body forces cannot be easily extracted from experiment, especially when coordination numbers can differ from the crystalline environment owing to broken or "floating" bonds present in the melt.

In addition to avoiding the use of empirical forces, pseudopotential simulations can provide a "window" on the microscopic structure of the l-state. One can examine the microstructure, the dynamical and electronic properties, and the nature of the bonds in the liquid state at any instant in time. Here we apply this approach to representative examples of semiconductors.

19.2 Constructing Liquid Ensembles

In order to describe the behavior of a liquid, "supercells" are commonly used to replicate an ensemble for the liquid state [6,7]. Supercells are large cells, which are constructed with periodic boundary conditions. As the simulation evolves, atoms can diffuse seamlessly in and out of supercell. This procedure allows a finite sample to be examined without explicitly considering any surfaces. Of course, introducing such an artificial constraint can alter the dynamics of the system, e.g. it may affect the radial and angular distributions of the liquid state. However, the role of the periodic constraint on physical properties can be tested by increasing the size of the cell and confirming that the structural and dynamical properties are insensitive to the cell size or shape.

To initiate the simulation, atoms are placed within the supercell either at random or within a crystalline structure. Other than the ease of implementation, a crystalline environment has few advantages over a random one. A crystalline environment retains long range order, which is inappropriate for the liquid state. Once an initial configuration has been obtained, the system is heated to a high temperature to randomize the atomic positions. One might couple the system to a fictive heat bath at high temperatures and allow the atoms to move a considerable distance from the initial positions. After some time, the system is cooled, as slowly as is allowed by computational constraints. When the desired temperature is achieved, the system can be removed from the heat bath and appropriate ensemble averages can be obtained from following the nuclear equations of motion. A typical thermalizing procedure is illustrated in Fig. 19.1.

The liquid temperature can be controlled by a variety of simulation techniques. These include the use of a Nose thermostat [8], or the use of Langevin dynamics [9,10]. There is a "procedural issue" as to the choice of thermostats. Does one suppress naturally occurring fluctuations in a finite system in order to preserve the desired temperature? Or does one fix a temperature that is

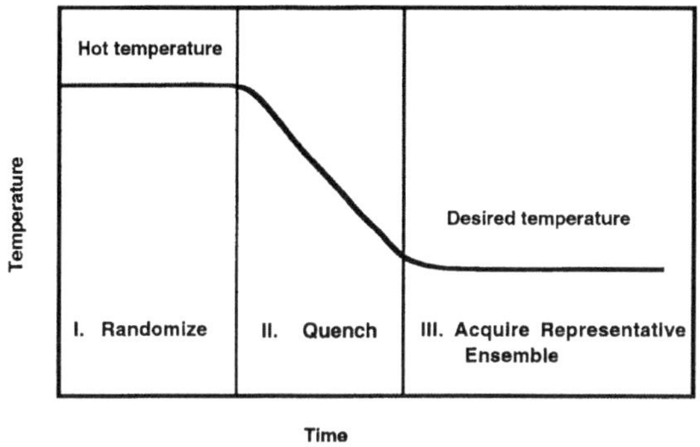

Fig. 19.1. Schematic schedule for preparing and simulating a liquid state ensemble

only approximately correct and allow fluctuations to occur as to be expected for a small system? The former procedure gives better control over the temperature, whereas the latter is more physical. We focus on the latter and employ Langevin dynamics. Within this description, the trajectory of each atom is computed from:

$$M_i \frac{d\boldsymbol{v}_i}{dt} = -\gamma M_i \boldsymbol{v}_i + \boldsymbol{G}_i(\gamma, T) + \boldsymbol{F}_i, \tag{19.1}$$

where $\boldsymbol{F}_i$ is the interatomic force on ith particle, M_i is the mass of the particle, and γ is the viscosity of the medium. The particles are subject to rapidly varying random forces, $\boldsymbol{G}_i(\gamma, T)$. Random forces and viscosity couple the system with the hypothetical heat bath. The interatomic forces, $\boldsymbol{F}_i$, can be obtained from classical force fields or computed quantum mechanically from the Hellmann–Feynman theorem, if the wave functions and Hamiltonian are known [11].

As the system is quenched, Langevin dynamics extracts energy from the system until the desired temperature is achieved. Once the system is thermalized, the viscocity term can be zeroed and the system transformed from a canonical to a microcanonical ensemble. Within the microcanonical ensemble one can examine diffusion and other dynamical properties. The chief concern is the size of the ensemble, or the supercell. Provided the cell is sufficiently large, fluctuations around the desired temperature will be minimal.

With respect to some technical issues, the equations of motion are usually integerated using a Verlet or modified Verlet algorithm [12]. The time steps are taken so that energy is conserved when Newtonian dynamics are utilized. Typically, the error in the energy conservation is at least one to two orders of magnitude less than the fluctuations in the kinetic (or potential) energies during the simulation. A value for the visocity parameter might be 0.001 a.u. (atomic units $e = \hbar = m = 1$) and a time step of $\Delta t = 5 - 10$ fs depending on the atomic mass of the constituent species. Simulation times can vary, but normally a minimum of $\sim 10^3$ time steps is employed, i.e. a simulation run in real time is usually 5–10 ps. Of course, classical simulations for liquids can be much longer than this, e.g., ~ 1 ns. The initial temperature in Fig. 19.1 is often taken to be $\sim 5,000$ K. The target temperature in all the simulations is slightly above the known melting points.

19.3 Quantum Interatomic Forces in the Liquid State

Simulation of liquids were commonly performed by utilizing classical potentials before efficient methods were developed for calculating quantum forces. A key step in this evolution was the application and extension of the pseudopotential method to molecular dynamics simulations. Car and Parrinello (CP) were among the first workers to perform ab inito molecular dynamics for liquids and other systems [13]. In the CP method, the nuclear degrees of freedom are handled classically and the electronic degrees of freedom are treated via fictive electronic dynamics.

An alternative path for performing simulations is to treat the electronic degrees of freedom by *real* dynamics, i.e., to consider the quantum forces between particles at each time step in the simulation. In this procedure, sometimes known as Born–Oppenheimer molecular dynamics (BOMD), the system is quenched at every time step to the Born–Oppenheimer surface and the electronic part of the problem solved via a self-consistent field [14]. One might assume that performing a fully self-consistent calculation at every step would be computationally intensive. However, because the forces are real, one can take signficantly larger time steps than within the CP method. The longer time step easily compensates for the extra computational load in obtaining a full self-consistent solution, e.g., often BOMD can have time steps ~ 100 times longer than in CP dynamics. Other advantages of BOMD include ease of implementation and handling metallic systems. This latter issue is straightforward. At every time time step, the levels are occupied as for a static system. One can also utilize level broadening to avoid problems that occur when level crossings occur.

Within the pseudopotential method, one needs to contruct an accurate pseudopotential. A wide variety of prescriptions have been devised to transcribe the all electron potential to a potential that replicates only the properties of the valence electrons [6,15]. A popular version of ab initio pseudopotentials has been proposed by Troullier and Martins [17]. The ion core pseudopotentials are determined by inverting the Kohn–Sham equations for an atom. The ion core potential is transferrable to other configurations as the potential corresponds to the nuclear and "inert" core electrons, both of which are not strongly dependent on the bonding environment. As an example, in silicon the 3 s and 3 p states would be treated as valence states and the ion core ($1\,s^2 2\,s^2 2\,p^6$) states are taken to be chemically inert. The ion core pseudopotential corresponds to a Si^{+4} configuration. Details of the pseudopotential construction can be found elsewhere [6].

Once the pseudopotential has been determined, the resulting eigenvalue problem needs to be solved for the system of interest:

$$\left(\frac{-\hbar^2\nabla^2}{2m} + V^p_{ion}(\boldsymbol{r}) + V_H(\boldsymbol{r}) + V_{xc}[\boldsymbol{r},\rho(\boldsymbol{r})] \right) \phi_n(\boldsymbol{r}) = E_n\phi_n(\boldsymbol{r}) \qquad (19.2)$$

where V^p_{ion} is the ionic pseudopotential for the system. Since the ion cores can be considered chemically inert and highly localized, it is a simple matter to write:

$$V^p_{ion}(\boldsymbol{r}) = \sum_{\boldsymbol{R}_a} V^p_{ion,a}(\boldsymbol{r} - \boldsymbol{R}_a) \qquad (19.3)$$

where $V^p_{\text{ion,a}}$ is the ion core pseudopotential associated with the atom, a, at a position $\boldsymbol{R}_\text{a}$. The ion core pseudopotential can contain nonlocal terms, but they are easily handled [6].

Another complexity of note concerns liquid II–VI semiconductors such as ZnTe. The 3d-states cannot be treated as core states in the pseudopotential construction. If one does so, poor structural properties will occur, e.g., the bond length in crystalline ZnTe might be $\sim$ 5–10% too short compared to experiment. Incorrect structural properties can have negative consequences in terms of computing radial distribution functions or diffusion constants. If the 3d-states are included as valence states, the structural properties are correctly reproduced. However, the computational load increases dramatically. The number of eigenvalues to be determined more than doubles and the wave functions highly localized, which complicates the choice of basis functions. Fortunately, one can avoid this problem by employing a "partial core" correction [16]. Partial core methods treat freeze the core state and do not treat it as part of the valence shell, but include an explicit treatment of the core-valence exchange-correlation potential. Partial core pseudopotential have been successfully tested by comparing three types of pseudopotentials: one with the 3d-states treated as part of the valence shell, one with the 3d-states treated as part of the core shell and one with the 3d-states treated via a partial core. Here we employ the partial core pseudopotentials for Zn. The resulting pseudopotential can be found in the literature.

The pseudopotentials we employ are "weak" and simple basis sets such as a plane wave basis can be quite effective for crystalline matter. For example, in the case of silicon only 50–100 plane waves per atom are required for well converged wave functions and eigenvalues [6]. The resulting matrix representation of the Schrödinger operator is dense in Fourier (plane wave) space, but it need not be formed explicitly [17,18]. The plane wave method uses a basis of the form:

$$\psi_{n,\boldsymbol{k}}(\boldsymbol{r}) = \sum_{\boldsymbol{G}} \alpha_n(\boldsymbol{k},\boldsymbol{G}) \exp(i(\boldsymbol{k}+\boldsymbol{G})\cdot\boldsymbol{r}) \qquad (19.4)$$

where $\boldsymbol{k}$ is the wave vector for the nth band state, $\boldsymbol{G}$ is a reciprocal lattice vector and $\alpha_\text{n}(\boldsymbol{k},\boldsymbol{G})$ represent the coefficients of the basis. For periodic systems, $\boldsymbol{k}$ is a good quantum number and the Bloch form (19.4) holds. If supercells are used, one still has a periodic system. However, $\boldsymbol{k}$, has no physical meaning other than reflecting the artificial periodicity of the system.

A variety of methods have been developed to find the eigenvalues/eigenvector pairs for the secular equations using a plane wave basis. These methods extract the pairs using interative solutions, e.g., generalized Davidson methods. Typically, the matrix size is less than $10,000 \times 10,000$. The cut-off for our plane wave basis is usually between 10–15 Ry. Details can be found in the literature [18].

Once the wave functions have been determined, the forces can be determined via the Hellmann–Feynman theorem [19]. These can be used in the dynamical description of the liquid, e.g., in the Langevin simulations. The simulations that incorporate these quantum forces are often called "ab initio simulations." This is in contrast to simulations that use interatomic force fields fit to classical interatomic potentials. For example, the very popular interatomic potentials for silicon such as the Stillinger–Weber potentials [20] are fit to experimental data bases and cannot be called first principles or ab initio potentials.

19.4 Properties of Liquids from Ab Initio Simulations

19.4.1 Microstructure of Liquids

With the possible exception of thermodynamic properties, the microstructure of liquids is the most fundamental materials property. If the structural properties of a system are unknown, it is virtually impossible to predict or understand other properties. Unfortunately, no experimental techniques are available for probing the atomistic structure of the liquid, save in an average sense. Both X-rays and neutron scattering can extract a structure factor, $S(q)$, for the liquid state [1]. The structure factor for an elemental liquid can be expressed as

$$S(q) = 1 + \frac{4\pi\rho_o}{q}\int_0^\infty [g(r) - 1] \; \sin(qr) r \, dr \tag{19.5}$$

where ρ_o is the macroscopic density of the liquid and $g(r)$ is the radial distribution or pair correlation function.

Determining the radial distribution function is complicated by the use of supercells. The value of $g(r)$ becomes unphysical for large r owing to the artificial periodicity of the supercell. For example, if one considers a cubic cell of size a, then for $r > 2a$ the positions of the atoms are precisely correlated.

For a compound liquid, the definition of the structure factor is modified. In particular, if one has two species (α, β), the structure factor can be written as

$$S_{\alpha\beta}(q) = \delta_{\alpha\beta} + \frac{4\pi\rho_{\alpha\beta}}{q}\int_0^\infty [g_{\alpha\beta}(r) - 1] \sin(qr) r \, dr, \tag{19.6}$$

where $\rho_{\alpha\beta}$ is a partial density. Unlike theoretical simulations, experimental probes cannot directly separate the contributions of the partial components. From theoretical work, it is possible to find the density of only the α species or the radial distribution of the β atoms around an α atom. If one knows the appropriate scattering parameters, one can compare calculations with neutron or X-ray scattering experiments. For example, the total structure function can be expressed as a linear combination of the partial structure factors, $S_{\alpha\alpha}(q)$, $S_{\alpha\beta}(q)$ and $S_{\beta\beta}(q)$, normalized by the scattering lengths (l_α, l_β) [21]

$$S(q) = \frac{{l_\alpha}^2 S_{\alpha\alpha}(q) + 2 l_\alpha l_\beta S_{\alpha\beta}(q) + {l_\beta}^2 S_{\beta\beta}(q)}{{l_\alpha}^2 + {l_\beta}^2}, \tag{19.7}$$

to find a value of $S(q)$, one needs only the ratio of the scattering lengths: l_α / l_β. The partial structure factors are obtained from partial radial distribution functions via a Fourier transformation.

19.4.2 Electronic Properties of the Liquid

Once we know the eigenpairs for a given atomic configuration are obtained, we can calculate the optical conductivity. According to Kubo–Greenwood expression [22] the real part of the conductivity can be expressed as:

$$\sigma_r(\omega) = \frac{2\pi e^2}{3m^2\omega\Omega} \sum_{n,m} \sum_{\alpha=x,y,z} |\langle \psi_m | p_\alpha | \psi_n \rangle|^2 \times \delta(E_n - E_m - \hbar\omega), \tag{19.8}$$

where E_i and ψ_i are eigenvalues and eigenfunctions, and Ω is the volume of the supercell. Dipole transition elements, $\langle \psi_m | p_\alpha | \psi_n \rangle$ are taken between the occupied and empty states. The function is sampled over a number of time steps as the simulation proceeds. Typically, the dielectric properties are converged once a dozen or time steps are sampled.

There are some issues associated with the Kubo–Greenwood expression. For example, the energy differences used evaluating the transitions are often taken to be the Kohn–Sham eigenvalues. It is well known in density functional theory that these transition energies do not represent the true optical transitions; the Kohn–Sham gaps tend to be too small resulting in the so-called "band gap" problem [23]. Another issue concerns extrinsic contributions to the conductivity. No impurity scattering is included. Despite these limitations, this formalism gives correct trends and often gives semi-quantitative agreement with experiment.

19.4.3 Dynamical Properties of Liquids: Diffusion

Diffusion constants are important parameters for characterizing the liquid and as input into modeling heat and mass flow in crystal growth simulations.

However, these physical parameters are difficult to measure. The diffusion constant can be determined from [24]

$$D = \lim_{t\to\infty} \frac{\langle [R(t)]^2 \rangle}{6t} \tag{19.9}$$

where

$$\langle [R(t)]^2 \rangle = \frac{1}{N} \sum_{i=1}^{N} |\boldsymbol{R}_i(t) - \boldsymbol{R}_i(0)|^2 \tag{19.10}$$

and where $\boldsymbol{R}_\mathrm{i}(t)$ is the position of the i th atom at time t. For compound semiconductors, a diffusion constant can be defined for each species, e.g., the diffusion of Ga in liquid GaAs can be obtained by examining the positions of the Ga atoms in the melt as a function of time.

19.5 Semiconductor Liquids: Si, GaAs and ZnTe

It is instructive to examine some prototypical semiconductor materials. Here we focus on three prototypical semiconductors: an elemental Group IV semiconductor (Si), a III–V semiconductor (GaAs) and a II–VI semiconductor (ZnTe). We examine the liquid state near the melting point using the thermal scheduling in Fig. 19.1.

In Fig. 19.2, we present the calculated structure factors and compare to experiment. The theoretical results were obtained using (19.5), i.e., the radial distribution function was calculated from an ensemble average and a Fourier transform performed. The experimental results were obtained from neutron measurements [25–27].

Several trends are apparent as one progresses from an elemental semiconductor (Si) to a fairly ionic semiconductor (ZnTe). For example, the line shape of the first spectral feature is remarkably different. In Si, there is predominant peak (near 2.3 Å^{-1}) with an accompanying shoulder. In ZnTe, the maximum of the structure factor (near 3.0 Å^{-1}) is preceded by a smaller peak. This can be attributed to strong differences in the microstructure of the melt, e.g., one can estimate the average coordination number of a liquid from the radial distribution function:

$$N_c = \int_0^{R_{\mathrm{min}}} 4\pi g(r) r^2 \, dr \,. \tag{19.11}$$

R_{min} is the first local minimum in the radial distribution function. In the crystalline state, both semiconductors would have a value of $N_c = 4$. This is not the case for the liquid state where $N_c \approx 6.4$ for Si and $N_c \approx 4.6$ for ZnTe. The structure factor of GaAs is qualitatively closer to Si than ZnTe; this is consistent its value of $N_c \approx 6.3$.

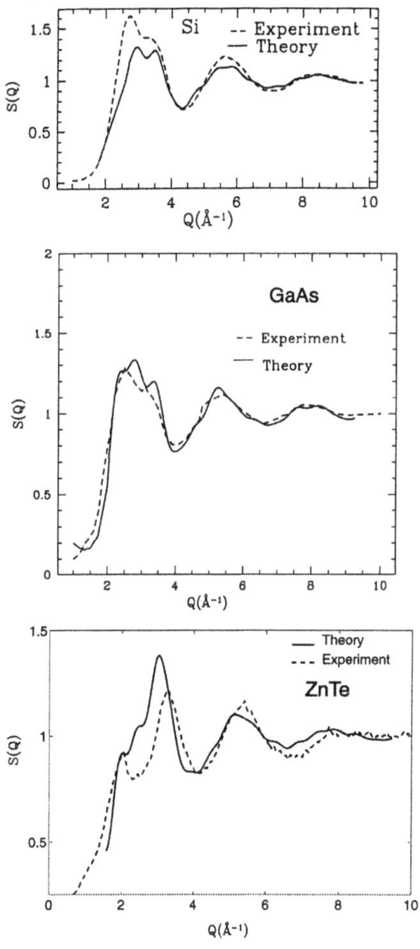

Fig. 19.2. Structure factors for Si, GaAs and ZnTe. The theoretical curves are from ab initio pseudopotential calculations. The experimental results are from [25–27] for Si, GaAs and ZnTe, respectively

In the case of heteropolar semiconductor, one can define a partial coordination number dependent on the particular species of interest:

$$N_{\alpha\beta} = \int_0^{R_{\min}} 4\pi g_{\alpha\beta}(r) r^2 \, dr \, . \tag{19.12}$$

Using this definition, one can extract a compositional disorder number, CDN:

$$CDN = (N_{\alpha\alpha} + N_{\beta\beta})/2N_{\alpha\beta} \, . \tag{19.13}$$

If $CDN = 0$, then no like-atom bonds occur as would be the case in the crystal. A value of $CDN = 1$ is consistent with a random mix and a value of

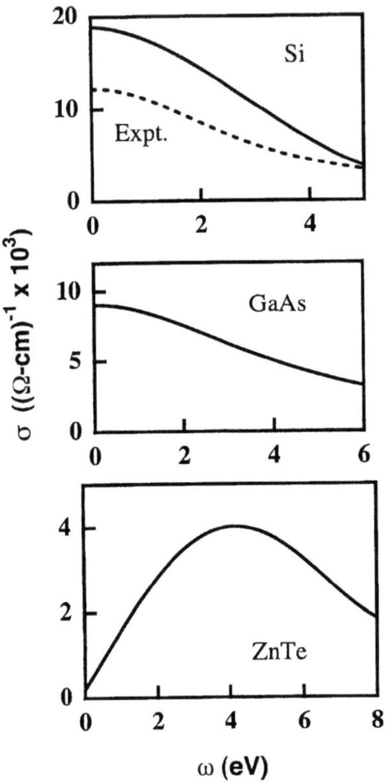

Fig. 19.3. Frequency dependent conductivity for Si, GaAs and ZnTe. The theoretical results for Si are after Stich et al. [28]. The experiment is from [29]

$CDN > 1$ indicates an association of cation–cation and anion–anion bonds. We obtain a value of $CDN = 0.8$ for GaAs and $CDN = 0.4$ for ZnTe. As such, one would expect l-GaAs to be characterized by more cation–cation and anion–anion bonds than l-ZnTe, but in both cases like atom bonds are less frequent than cation–anion bonds.

The structural differences evident between Si, GaAs and ZnTe are also evident in their electronic properties. The results for Si are from Stich et al. [28]. Si and GaAs have similar coordination numbers for liquid microstructure. These microstructures are more "metallike" than ZnTe. This situation is reflected in the frequency dependent conductivities as indicted in Fig. 19.3.

Si is known to be a metal in the liquid state. Measurements of the Hall coefficients and electrical conductivity show that the conduction electron density increases by 10^{10} compared to the solid state [1]. While the electron density increases, the scattering rates increases in the melt owing to the absence of any long range order. The combination of these factors results in a conductivity increase by a factor of 20 from the solid state to liquid state [1]. If the

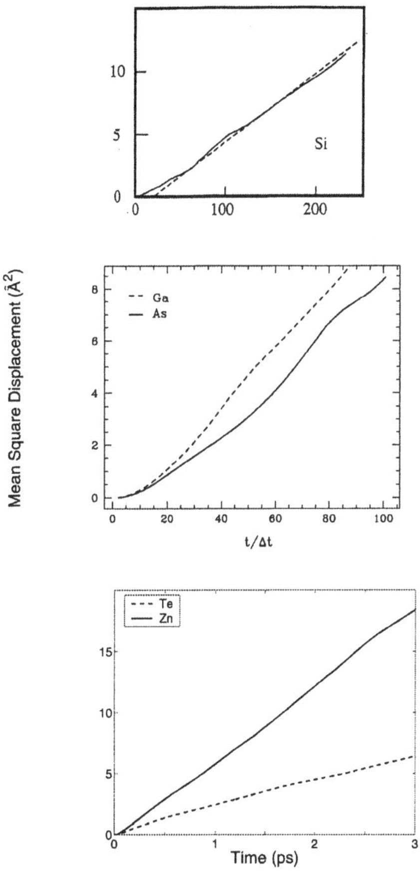

Fig. 19.4. The time dependence of the mean square displacement of atoms in the liquid state. The temperature of the liquid is fixed to be slightly above the melting point

temperature of the melt is increased, the conductivity *decreases* as expected for a metal. The conductivity behavior for GaAs is similar. Upon melting the conductivity increases by a factor 26 and decreases with increasing temperature [1]. The conductivity of *l*-ZnTe is significantly smaller than either *l*-Si or *l*-GaAs. For example, the measured dc-conductivity of *l*-Si is $\sim 10^4$ $(\Omega\text{-cm})^{-1}$ at the melting point; the conductivity of ZnTe near the melting point is $\sim 10^2$ $(\Omega\text{-cm})^{-1}$ [1]. Also, the conductivity of *l*-ZnTe *increases* with increasing temperature. This is characteristic of a semiconductor.

In general, elemental semiconductors like Si and Ge are metallic in the melt, as are III–V semiconductors. In contrast, most II–VI semiconductors are not. A notable exception is HgTe, which is not a semiconductor in the crystalline state, but a semi-metal.

Diffusion constants are difficult to measure, but maybe determined in a straightforward fashion by ab initio simulations. The root mean square displacement for atomic species in the melt are illustrated in Fig. 19.4. The slope of this curve gives the diffusion constant. Owing to the finite size of the supercell, some statistical noise is present, but the slopes are well converged. For Si, no experimental measurements exist for the self-diffusion. However, our calculated value of $D = 1.9 \times 10^{-4} \mathrm{cm}^2/\mathrm{sec}$ is consistent with other theoretical values [28]. In the case of GaAs, an average diffusion constant has been measured: $D = 1.6 \times 10^{4} \mathrm{cm}^2/\mathrm{sec}$ [30]. The simulated values are $D_{Ga} = 1.6 \times 10^{-4} \mathrm{cm}^2/\mathrm{sec}$ and $D_{As} = 1.2 \times 10^{-4} \mathrm{cm}^2/\mathrm{sec}$. This is in contrast to tight binding simulations. These simulations tend to underestimate diffusion constants: $D_{Ga} = 1.0 \times 10^{-4} \mathrm{cm}^2/\mathrm{sec}$ and $D_{As} = 0.7 \times 10^{-4} \mathrm{cm}^2/\mathrm{sec}$ [31].

Simulations based on approximate methods such as tight binding or interatomic potentials often underestimate the diffusion constants for liquids. It is likely that a restrictive description of the electronic degrees of freedom in the tight binding method or the complete neglect of electronic coordinates in classical simulation is the origin of this problem. In the case of ZnTe, we find $D_{Zn} = 1.0 \times 10^{-4} \mathrm{cm}^2/\mathrm{sec}$ and $D_{Te} = 0.32 \times 10^{-4} \mathrm{cm}^2/\mathrm{sec}$.

19.6 Summary

We have illustrated how one can simulate the properties of semiconductor liquids using quantum forces. In particular, we illustrated how the pseudopotential-density functional method can be used to perform simulations of a variety of semiconductor liquids. In contrast to classical simulations, which use empirical interatomic forces, quantum forces eliminate any ad hoc assumptions about the interatomic forces. For example, charge transfer, coordination changes and rehybridization are accurately reproduced using pseudopotentials. This eliminates any ambiguity arising from the transcription of classical interactions to quantum ones. In addition, the *electronic* states of the melt are known at each time step of the simulation. This allows one to examine electronic properties of the melt.

Simulations will play a prominent role in the study of liquids. Unlike solid state phases where the structures are fixed, and known for crystalline matter, liquids involve complex, dynamical interactions where the nature of the chemical bond is continually changing. We have provided some illustrations here of how this process can be performed for semiconductor liquids. Liquid simulations should help bring about insights and invaluable guidance as we pursue a deeper understanding of this state of matter.

Acknowledgments

We would like to acknowledge support for this work by NASA, the National Science Foundation, the U.S. Department of Energy and the Minnesota Supercomputer Institute.

References

1. V.M. Glazov and O.D. Shchelikov: **18**, 411 (1984); V.M. Glazov, S.N. Chizhevskaya and N.N. Glagoleva: *Liquid Semiconductors* (Plenum Press, 1969)
2. I. Stich, R. Car and M. Parrinello: Phys. Rev. Lett. **63**, 2240 (1989); Q.-M. Zhang, G. Chiarotti, A. Selloni, R. Car and M. Parrinello: Phys. Rev. B **42**, 5071 (1990); R.V. Kulkarni and D. Stroud: Phys. Rev. B **62**, 4991 (2000); R.V. Kulkarni and D. Stroud: Phys. Rev. B **55**, 6896 (1997)
3. J.R. Chelikowsky, J.J. Derby, V. Godlevsky, M. Jain and Y.R. Raty: J. Phys. Cond. Matt. **13**, R817 (2001)
4. G. Ciccotti, D. Frenkel, and I.R. McDonald (Eds.): *Simulation of Liquids and Solids* (North Holland, Amsterdam 1987)
5. J.R. Chelikowsky, K. Glassford, and J.C. Phillips: Phys. Rev. B **44**, 1538 (1991)
6. J.R. Chelikowsky, and M.L. Cohen: "Ab initio Pseudopotentials for Semiconductors," *Handbook on Semiconductors*, Editor: P. Landsberg (Elsevier, 1992) Vol. 1, p. 59; J.R. Chelikowsky and S.G. Louie (Eds.): *Quantum Theory of Real Materials*, (Kluwer Press, 1996); W.E. Pickett: Computer Physics Reports **9**, 115 (1989); J.R. Chelikowsky: J. Phys. D **33**, R33 (2000)
7. M.L. Cohen, M. Schlüter, J.R. Chelikowsky, and S.G. Louie: Phys. Rev. B **12**, 5575 (1975) and references therein
8. S. Nosé: Mol. Phys. **52**, 255 (1984); J. Chem. Phys. **81**, 511 (1984)
9. R. Kubo: Rep. Prog. Theor. Phys. **29**, 255 (1966); H. Risken, *The Fokker-Planck Equation* (Springer, Berlin Heidelberg New York 1984); R.L. Stratanovitch: *Topics in the Theory of Random Noise* (Gordon Breach, New York 1967); N.G. van Kampen: *Stochastic Processes in Physics and Chemistry* (North Holland, Amsterdam 1981)
10. Examples of Langevin applications include: J.C. Tully, G. Gilmer, and M. Shugart: J. Chem. Phys. **71**, 1630 (1979); R. Biswas and D.R. Hamann: Phys. Rev. B **34**, 895 (1986); N. Binggeli and J.R. Chelikowsky: Phys. Rev. B **50**, 11764 (1994)
11. N. Binggeli, J.L. Martins, and J.R. Chelikowsky: Phys. Rev. Lett. **68**, 2956 (1992)
12. L. Verlet: Phys. Rev. **165**, 201 (1967); D. Beeman: J. Comp. Phys. **20**, 130 (1976)
13. R. Car and M. Parrinello: Phys. Rev. Lett. **55**, 2471 (1985); **60** 204 (1988). A recent review of the Car-Parrinello and related methods can be found in D. Marx and J. Hutter: *Ab Initio Molecular Dynamics: Theory and Implementation*, in *Modern Methods and Algorithms of Quantum Chemistry*, Editor: J. Grotendorst (John von Neumann Institute for Computing, Forschungszentrum Jülich 2000) pp. 301–449
14. R.M. Wentzcovitch and J.L. Martins: Solid State Comm. **78**, 831 (1991); N. Binggeli, J.L. Martins, and J.R. Chelikowsky: Phys. Rev. Lett. **68**, 2956 (1992); J.R. Chelikowsky and N. Binggeli: Solid State Comm. **88**, 381 (1993); Phys. Rev. B **49**, 114 (1994)
15. D.R. Hamann, M. Schlüter, and C. Chiang: Phys. Rev. Lett. **43**, 1494 (1979); G. Kerker: J. Phys. C **13**, L189, (1980); D. Vanderbilt: Phys. Rev. B **32**, 8412, (1985); A. Rappe, K.M. Rabe, E. Kaxiras, and J.D. Joannopoulos: Phys. Rev. B **43**, 1227 (1990); M.L. Cohen and J.R. Chelikowsky: *Electronic Structure*

and Optical Properties of Semiconductors, Springer Solid-State Science Vol. 75, (Springer, Berlin Heidelberg New York 1988)
16. S. Louie, S. Froyen, and M. Cohen: Phys. Rev. B **26**, 1738 (1982)
17. N. Troullier and J.L. Martins: Phys. Rev. B **43**, 1993 (1991)
18. Y. Saad, A. Stathopoulos, J.R. Chelikowsky, K. Wu, and S. Öğüt: BIT **36**, 563 (1996); A. Stathopoulos, S. Öğüt, Y. Saad, J.R. Chelikowsky and H. Kim: Computing in Science and Engineering **2**, 19 (2000); S.G. Louie, in *Electronic Structure, Dynamics and Quantum Structural Properties of Condensed Matter*, (Plenum, New York 1985) p. 335
19. C.T. Chan, K.P. Bohnen, and K.M. Ho: Phys. Rev. **B47**, 4771 (1993)
20. J.Q. Broughton and X.P. Li: Phys. Rev. B **35**, 9120 (1987); F.H. Stillinger and T.A. Weber: Phys. Rev. B **31**, 5262 (1985); M.D. Kluge, J.D. Ray, and A. Rahman: Phys. Rev. B **36**, 4234 (1987); D. Leudtke and U. Landman: Phys. Rev. B **40**, 1164 (1989)
21. J. Emsley: *The Elements*, 3rd edn. (Oxford University Press, Oxford 1996)
22. R. Kubo: J. Phys. Soc. Jpn. **12**, 570 (1957); D. Greenwood: Proc. Phys. Soc. **71**, 585 (1958)
23. M.S. Hybertsen and S.G. Louie: Phys. Rev. B **34**, 5390 (1986)
24. R. Virkkunen, K. Laasonen, and R. Nieminen: J. Phys. Condes. Matter **3**, 7455 (1991)
25. Y. Waseda: *The Structure of Non-crystalline Materials* (McGraw Hill, New York 1980)
26. C. Bergman, C. Bichara, P. Chieux and J. Gaspard: J. Physique Col. **C8–46**, 97 (1985)
27. J. Gaspard, J. Raty, R. Ceolin, and R. Bellissent: J. Non-Cryst. Solids **205–207**, 75 (1996)
28. I. Stich, R. Car and M. Parrinello: Phys. Rev. B **44**, 4262 (1991)
29. K.M. Shvarev, B.A. Baum, and P.V. Geld: Sov. Phys. Solid State **16**, 2111 (1975)
30. K. Hellwege (Ed.): *Landölt-Bornstein New Series* (Springer, Berlin Heidelberg New York 1984)
31. C. Molteni, L. Colombo, and L. Miglio: J. Phys. Cond. Matter **2**, 279 (1994)

20 Mapping Molecular Simulations to Continuum Models for Binary Fluids

C. Denniston and M.O. Robbins

Dept. of Physics and Astronomy, Johns Hopkins University
Baltimore, MD 21218, USA.

Abstract. In this paper we show recent results on how molecular dynamics simulations can be used to quantitatively map out a mesoscopic continuum model for binary fluids. We map not only the order parameter profile, but also the density and the stress. These are essential details in order to be able to fully describe the dynamics. The boundary conditions for a miscible binary fluid are also described. The molecular dynamics reveals unexpected terms in the boundary conditions, information which cannot be obtained by any other method.

20.1 Introduction

There is increasing interest in using molecular dynamics simulations to map out not just parameters from a well-defined continuum model, but also to aid in the development of the continuum model itself. Part of the difficulty in accomplishing this task is that even with modern computers and algorithms, it is still not possible to do a molecular dynamics simulation on the length and time scales typically associated with macroscopic continuum models. As a result, one must use an intermediate mesoscopic continuum model. While a continuum model may apply down to the mesoscale, many of the assumptions made in macroscopic models, such as the stress tensor being a purely local function or incompressibility for fluids, may be violated.

In this paper we describe some of the recent progress in performing this mapping for binary fluids. The next section describes our molecular dynamics simulations. In the third section we review details for measuring the microscopic stress tensor and how such measurements can be affected by finite size effects. The fourth section demonstrates how the measured stress, density and order parameter profile for a liquid–liquid interface can be quantitatively fit using a square gradient theory that accounts for variations in both the density and order parameter. Any continuum model needs boundary conditions at a solid surface, and these are discussed for miscible binary fluids in the fifth section.

20.2 Simulations and Microscopic Models

We consider a mixture of two types of molecules, labeled a and b. The interactions between atoms of type i and j separated by a distance r are modeled using a Lennard–Jones (LJ) potential,

$$V_{ij}(r) = 4\epsilon_{ij}\left[(\sigma_{ij}/r)^{12} - (\sigma_{ij}/r)^{6}\right], \tag{20.1}$$

where ϵ_{ij} specifies the interaction energy and σ_{ij} the interaction length. Their averages are denoted by ϵ and σ, respectively. A characteristic time scale is given by $\tau = \sigma(m/\epsilon)^{1/2}$, where m is the average of the molecular masses m_i.

For the miscible fluids studied in Sect. 20.5, the force is truncated at $r_c = 2.2\sigma$ to improve computational efficiency. For the partially miscible fluids examined in Sect. 20.4, we cut of the potential at $r_c = 2^{1/6}\sigma$ so that the interactions are purely repulsive. Two molecules of the same type interact with $\epsilon_{aa} = \epsilon_{bb} = \epsilon$. An extra repulsion is added between unlike molecules $\epsilon_{ab} = \epsilon_{ba} = \epsilon(1+\epsilon^*)$. By varying ϵ^* we can vary the miscibility of the two fluids from completely miscible to completely immiscible. Similar models have been used by workers who have studied phase diagrams [1,2], surface tensions [3], slip [4] and critical phenomena of polymer fluids [1]. Ideally, we would also like to examine polymer mixtures, however we are interested in convection induced by chemical potential (or concentration) gradients. Such processes take much longer to develop in polymer fluids so we will restrict our attention here to monomers.

The equations of motion for molecule i located at position $\boldsymbol{r}_i = (x_i, y_i, z_i)$ are

$$\begin{aligned} m_i\frac{d^2x_i}{dt^2} &= -\frac{\partial U}{\partial x_i}\,, \\ m_i\frac{d^2y_i}{dt^2} &= -\frac{\partial U}{\partial y_i} - m\Gamma\frac{dy_i}{dt} + W_i(t)\,, \\ m_i\frac{d^2z_i}{dt^2} &= -\frac{\partial U}{\partial z_i}\,. \end{aligned} \tag{20.2}$$

The x and z equations are just Newton's equations of motion for a particle in a potential $U = \sum_{i,j} V_{ij}$. The motion in the y direction is coupled to a heat bath, acting through a weak stochastic force $W_i(t)$ and a corresponding damping term with friction coefficient Γ. In all the systems we will study, we have periodic boundary conditions in the y direction and any averaged quantity is translationally invariant in the y direction, or zero in the case of velocities. However, average quantities do exhibit nonvanishing gradients and velocities in the x and z directions due to the presence of walls, interfaces, or concentration gradients. By coupling the heat bath to only the y direction, we can preserve momentum conservation in the x and z directions. The temperature, which is related to the second moment of the velocity distribution remains isotropic as long as the stochastic force is weak, and this is easily verified in the simulations. We use $\Gamma = 0.5\tau^{-1}$ which corresponds to underdamped motion so that even in the y direction the motion is dominated by inertia.

The equations of motion are integrated using a velocity-Verlet algorithm [5] in the LAMMPS package [6]. LAMMPS is a classical molecular dynamics code designed for simulating molecular and atomic systems on parallel computers using spatial-decomposition techniques. It runs on any parallel platform that supports F90 and the MPI message-passing library or on single-processor workstations. We run it on both an Intel Pentium III based PC cluster and on an AMD Athlon based cluster.

In addition to the base LAMMPS package we have added a number of routines for our specific purposes. These include a grand canonical Monte Carlo equilibration routine. To speed up the equilibration time, we supplement the molecular dynamics moves with a Monte Carlo procedure relabeling the molecules (from A to B or visa versa) according to the Metropolis transition rule. This is a procedure first used for lattice simulations [1] and latter extended to molecular dynamics simulations of polymer mixtures [2]. After equilibration, the Monte Carlo routines are turned off. A substantial time period may still separate the apparent stabilization of the density and order parameter profiles and when any steady-state convective flow is developed. In addition to ensuring an accurate and efficient means of simulating the equations of motion, considerable care also needs to be taken in measuring quantities like the stress, as described in the next section.

20.3 Microscopic and Macroscopic Stress

The macroscopic stress can be obtained easily from a sum over the whole system of

$$P_{\alpha\beta} = \frac{1}{V}\left(\sum_i m_i v_{i\alpha} v_{i\beta} - \sum_i \sum_{j>i} r_{ij\alpha} \frac{\partial V_{ij}}{\partial r_{i\beta}}\right), \tag{20.3}$$

where $\boldsymbol{r}_{ij}$ is the vector between particles i and j. However, as we are interested in mapping our molecular simulations onto a continuum at the finest scale at which this makes sense (i.e. at distances just large enough that the discrete nature of the particles has been averaged out), we are interested in a locally defined stress. The difficulty in defining a local stress is related to an apparent ambiguity about where in space to assign the contribution in the second term of (20.3) (the interaction term). The local pressure tensor has the following form [5]

$$P_{\alpha\beta}(\boldsymbol{r}) = k_B T \rho(\boldsymbol{r}) - \sum_i \sum_{j>i} \left\langle \frac{r_{ij\alpha}}{r_{ij}} \frac{\partial V_{ij}}{\partial r_{ij}} \int_{C_{ij}} dr'_\beta \, \delta(\boldsymbol{r} - \boldsymbol{r}') \right\rangle$$

$$= k_B T \rho(\boldsymbol{r}) + \sum_i \sum_{j>i} \left\langle f_{ij\alpha} \int_{C_{ij}} dr'_\beta \, \delta(\boldsymbol{r} - \boldsymbol{r}') \right\rangle . \qquad (20.4)$$

The apparent ambiguity arises due to the fact than *any* contour C_{ij} joining atom i and j would appear to be acceptable (in that it will satisfy microscopic momentum balance). However, if one requires that the microscopic stress also conform to symmetry and transformation properties obeyed by the macroscopic stress the contour choice becomes unique. This is true whether or not you impose these additional requirements on the scale of the averages [7,8] in (20.4) or on the corresponding microscopic instantaneous many-body variable inside the averages [9]. The appropriate contour is just the straight line between the two atoms, a choice which was originally proposed by Irving and Kirkwood [7].

Obviously, one cannot accumulate sufficient data to ascertain $P_{\alpha\beta}$ at every point in space. One must do some binning of the data onto a grid. This can be computationally costly as the binning must be done for every force pair in the system. As a result it can easily be just as time consuming as integrating the equations of motion (20.2) themselves. A number of possible simplifications to reduce computational load have been suggested. One in fairly common use is to just associate $f_{ij\alpha}$ with the atoms at the ends, rather than distributing it along the whole line between them. The $f_{ij\alpha}$ associated with each atom is then accumulated and binned periodically, rather than after every calculation of a force. Unfortunately, if there are any density gradients in the system this can result in unphysical effects. For example, one may obtain variations in the stress component normal to a static liquid–liquid interface that imply mechanical instability. The only simplification that does not lead to unphysical effects is to not do the stress calculation after every time step, but only after sufficient number of steps (~ 10) so that the stress measurements are somewhat uncorrelated.

Once one has calculated the stress, the surface tension between the two immiscible fluids can be computed from the mechanical definition due to Kirkwood and Buff [10]. The surface tension is given by integrating the difference between the normal and parallel components of the pressure across the interface,

$$\gamma = \int [P_\perp(r) - P_\parallel(r)] \, dr . \qquad (20.5)$$

Far from the interface the pressure is isotropic and $P_\perp = P_\parallel$. For a flat interface normal to the x-axis (20.5) is

$$\gamma = \int [P_{xx}(r) - (P_{yy}(r) + P_{zz}(r))/2] \, dx . \qquad (20.6)$$

In the quiescent state P_{xx} is constant (no flow implies $\partial_\alpha P_{\alpha\beta} = 0$ and there is no variation in either the y or z direction) throughout the system. This clearly implies that if we can quantitatively understand the stress profile through the interface, things like the surface tension are guaranteed to work out correctly. This sort of detailed examination of the stress profile (as opposed to its integral, the surface tension) has not been done, presumably due to the (incorrectly) perceived ambiguity in the definition of the microscopic stress.

Typically what has been done is to average the stress over the whole system, (20.3) and then take

$$\gamma = (P_\perp - P_\|)L_\perp \,, \tag{20.7}$$

where $L_\perp$ is the size of the simulation cell in the direction perpendicular to the interface. It has been noticed by a number of authors that comparisons of the surface tension measured in this way to other techniques tend to show small systematic differences [3,5,11]. While different possible explanations have been put forward, the true reason is most likely finite size effects which are missed by using the globally averaged stress rather than examining the local stress.

This can be most easily seen by examining the fairly simple geometry shown in Fig. 20.1. There are periodic boundary conditions in all directions and two flat (on average) interfaces normal to the x-axis. The average density is $\rho = 0.85\,\mathrm{m}/\sigma^3$, the interactions are purely repulsive ($r_c = 1.1\sigma$), and $\epsilon_{aa} = \epsilon_{bb}$. Phase separation is driven by the fact that $\epsilon_{ab} = 7\epsilon_{aa}$. Direct use of (20.7) gives $\gamma = 0.275 \pm 0.005$. However, a calculation of $(P_\perp - P_\|)L_\perp$ for a similar system *without* interfaces gives $(P_\perp - P_\|)L_\perp = 0.15 \pm 0.01$, i.e. there is an anisotropy in the pressure tensor even in the bulk regions. An examination of the microscopic pressure tensor in the bulk regions (away from the interfaces where it should be hydrostatic) of the simulation shown in Fig. 20.1 or in the system without interfaces shows that $(P_\perp - P_\|) = 0.004$.

A system twice as large (i.e. $L_y = L_z = 16.4\sigma$ and $L_x = 75.6\sigma$) but otherwise identical gives $(P_\perp - P_\|) \leq 0.0002$. Thus the larger system is isotropic to within measurement error whereas the smaller system has a small but very significant anisotropy which leads to an error in the surface tension which in this case is of order 100%. For the more typically studied completely immiscible systems where $\gamma \sim 1$ this systematic error will be smaller ($\sim 15\%$) but still significant. The origin of this finite size effect is fairly simple: Although the smaller system is larger than the correlation length (which is about 2.5σ here and for other typical Lennard–Jones fluids) the correlation function is

Fig. 20.1. A liquid–liquid interface between Lennard–Jones fluids. System size is $L_y = L_z = 8.2\sigma$ and $L_x = 37.8\sigma$

still significantly different from zero at distances of the order of the transverse size (along y and z, but is essentially zero for distances of order of the longitudinal size). This means that molecules can "feel" themselves in the transverse direction and leads to the observed anisotropy.

Information gained from the actual microscopic stress profile through the liquid–liquid interface can also give valuable insights into the structure of the free energy barrier between the phases in the system. This will be described in the next section.

20.4 Interface Stresses

To avoid the finite size effects discussed in the previous section we examine only systems large enough so that in bulk regions the pressure is isotropic to well within our measurement error (≤ 0.0002). In particular we will use system sizes of $L_y = L_z = 16.1\sigma$ and $L_x = 74.4\sigma$. With periodic boundary conditions, the equilibrium configuration is to have two interfaces perpendicular to the x-axis, separating A rich and B rich phases. The profile of the order parameter, $\phi(x) = (\rho_A(x) - \rho_B(x))/\rho$, going through one of the interfaces is shown in Fig. 20.2. This is obtained by averaging over the y and z directions. The shape of the profile is well fit by the form $\phi(x) = \phi_0 \tanh(2x/\xi)$, where ξ defines the interface width. The other relevant parameter in the problem is the density ρ. The profile of the density through the interface is shown in Fig. 20.2. This density dip is a ubiquitous feature of fluid–fluid interfaces and has been observed in both simple fluids [12,13] and in immiscible polymer fluids [14]. The dip occurs because the energetically unfavorable a–b particle interactions are concentrated at the interface. As a result, the system can lower its energy by decreasing the overall density at the surface.

The measured profiles of the diagonal components of the pressure tensor are shown in Fig. 20.3. The P_{xx} component is flat through the interface, as it must be in order for the steady state condition $\nabla \cdot \boldsymbol{P} = 0$ to be satisfied. The components parallel to the interface P_{yy} and P_{zz} exhibit a dip at the interface. This dip is the mechanical manifestation of the surface tension. We now present a model capable of quantitatively fitting these profiles.

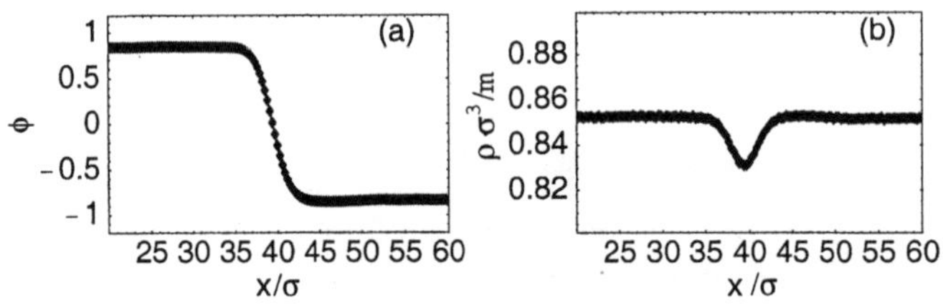

Fig. 20.2. The interfacial profiles of the yz-averaged (**a**) order parameter and (**b**) density for a liquid-liquid interface between Lennard–Jones fluids. System size is $L_y = L_z = 16.1\sigma$ and $L_x = 74.4\sigma$

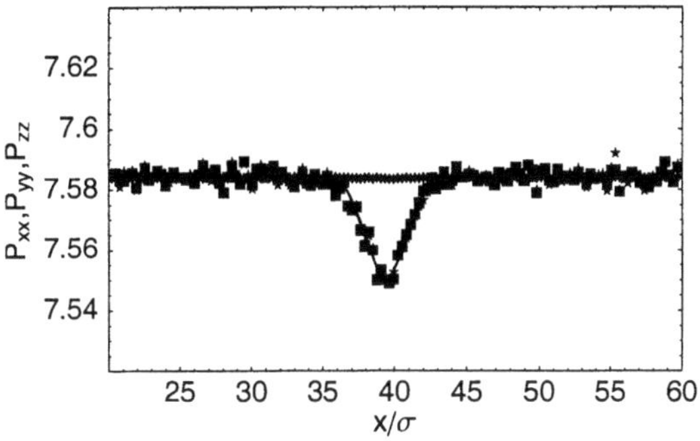

Fig. 20.3. The interfacial profiles of the yz-averaged diagonal components of the pressure tensor P_{xx} (diamonds on the flat line), P_{yy}, and P_{zz} (squares and stars on the line that exhibits a dip at the interface). The lines are fits to the square gradient theory described in the text

There are a number of possible models one could try to apply to this situation. Density functional theory can be quite reasonable at equilibrium interfacial properties [15]. However, we are most interested in using the model to study dynamics at long time and length scales, including hydrodynamic modes and convection-diffusion effects. As such, we will work with a simpler model, namely square gradient theories, which are easily simulated using mesoscopic techniques such as lattice Boltzmann algorithms.

We start with the free energy functional,

$$\mathcal{F} = k_B T \int dr \left\{ \psi(\rho, \phi, T) + \frac{1}{2} K_\rho (\nabla \rho)^2 + \frac{1}{2} K_\phi (\nabla \phi)^2 \right\} , \tag{20.8}$$

where K_ρ and K_ϕ are "elastic" constants, to be determined, and $\psi(\rho, \phi, T)$ is the bulk free energy, which determines the equilibrium state in gradient free regions. It can be shown [16] that this free energy implies a pressure tensor (non-dissipative stress) of the form

$$\begin{aligned} P_{\alpha\beta} = & -k_B T \left(\psi - \rho \frac{\partial \psi}{\partial \rho} + K_\rho \rho \nabla^2 \rho \right) \delta_{\alpha\beta} \\ & + k_B T K_\rho \left[(\partial_\alpha \rho)(\partial_\beta \rho) - \frac{1}{2} (\partial_\gamma \rho)^2 \delta_{\alpha\beta} \right] \\ & + k_B T K_\phi \left[(\partial_\alpha \phi)(\partial_\beta \phi) - \frac{1}{2} (\partial_\gamma \phi)^2 \delta_{\alpha\beta} \right] . \end{aligned} \tag{20.9}$$

It is now straightforward to show that the kernel of the surface tension integral (20.6) is given by

$$\Gamma(x) \equiv P_{xx}(x) - (P_{yy}(r) + P_{zz})/2 , \tag{20.10}$$
$$= k_B T K_\rho (\partial_x \rho)^2 + k_B T K_\phi (\partial_x \phi)^2 . \tag{20.11}$$

Thus we see that comparison of the measured Γ to the prediction of the square gradient theory requires a simple linear two parameter fit using the measured values of $\partial_x \rho$ and $\partial_x \phi$. This fit is shown as lines in Fig. 20.3. The fit is excellent and all residuals are zero to within the statistical precision of the data.

The inclusion of the square gradient of the density (and not just the order parameter) is imperative in obtaining quantitative agreement with the simulations. In this case, even though the density gradients are much smaller than the gradients in the order parameter, K_ρ is significantly larger than K_ϕ so that both terms contribute comparably to the free energy. We have also examined much sharper interfaces, including those where the dip in density is comparable to the change in the order parameter and still find that the square gradient theory describes the data very well [16]. A detailed description of these results, as well as those for the bulk free energy, is beyond the space allocated here.

20.5 Miscible Fluids: Boundary Stress Induced Flow

In addition to bulk properties and liquid–liquid interfaces, a complete quantitative mesoscopic description of binary fluids requires information about boundary conditions at a solid wall. We have examined this problem for miscible binary fluids [17,18].

We use the channel geometry shown in Fig. 20.4, which is similar to that used for a single fluid in [19]. There are two walls, at $z = 0$ and $z = L$, and periodic boundary conditions in the xy-plane. The walls contain type w atoms fixed to lattice sites forming two (111) layers of an fcc surface. To obtain a steady-state concentration, we construct a "Maxwell Demon's wall" at the end of the channel in the x-direction. The wall is less than one σ thick. When atoms of type a enter the wall region they are preferentially transmitted from left to right and conversely for atoms of type b (atoms going through the wall are transcribed periodically to the other end of the channel).

Fick's law relates the diffusive flux of particles of type j,

$$\boldsymbol{J}_j \equiv \rho_j(\boldsymbol{v}_j - \boldsymbol{v}) , \tag{20.12}$$

to the gradient of the concentration $c_j \equiv \rho_j/\rho$ [20],

$$\boldsymbol{J}_j = -\rho D \nabla c_j , \tag{20.13}$$

where D is the diffusion constant and $\boldsymbol{v} \equiv (\sum_i \rho_i \boldsymbol{v}_i)/\rho$ is the barycentric, or mass averaged, velocity. As expected, in the bulk regions we find that Fick's law is obeyed to within our measurement uncertainty.

The boundary conditions for simple single component LJ fluids are well described by the Navier slip condition [19,21]. This assumes that the velocity

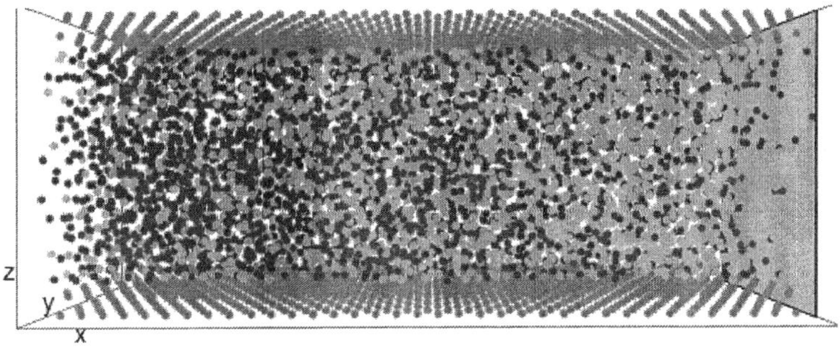

Fig. 20.4. Miscible binary mixture between two solid walls. The two walls, at $z = 0$ and $z = L_z$ contain atoms fixed to lattice sites forming tow (111) layers of an fcc surface. There are two types of atoms between the walls, differentiated by their shading (light and dark). There are periodic boundary conditions in the y direction and a periodic "Maxwell's demon" wall at ends of the channel in the x-direction (see text)

difference between solid and fluid is proportional to the viscous shear stress. For our geometry this implies

$$v_x|_{\text{wall}} = L_s \partial_z v_x|_{\text{wall}} , \tag{20.14}$$

where L_s is called the slip length. A larger value of L_s implies less drag at the interface. In most cases, the slip length has atomic dimensions [19,21]. For a fluid composed of one of the components we use here, the slip length is less than 2σ.

It had been suggested [22] that the boundary condition for a mixture of two fluids should be such that the individual velocities $\boldsymbol{v}_j$ of the fluid components should vanish at the wall in the case where the two fluids individually (i.e. in a 100% mixture) obeyed a "no-slip" boundary condition. It was subsequently pointed out [23] that this boundary condition contradicts Fick's law of molecular diffusion, at least in its usual form where the diffusion coefficient is assumed to be position independent (if D is constant in (20.13) then requiring all velocities to vanish at the wall would also require any concentration gradients to vanish at the wall, a requirement which over-specifies the boundary conditions). It turns out, however, that the Navier slip condition (which includes "no-slip" boundary conditions) is *not* the appropriate boundary condition for a miscible binary fluid.

This is most easily seen by examining a situation where the mass of the species is different. In particular we take the situation depicted in Fig. 20.4 with $m_a = 1.25\,\text{m}$ and $m_b = 0.75\,\text{m}$, with $a(b)$ referring to the darker (lighter) molecules in the figure. The total numbers of a and b atoms in the system are equal and the concentration gradient is extremal in the sense that at the left end of the channel the system is all a molecules and at the right end it is all b molecules. In addition to the diffusive flow, we can add a positive or negative pressure gradient down the channel by adding a small constant force on the

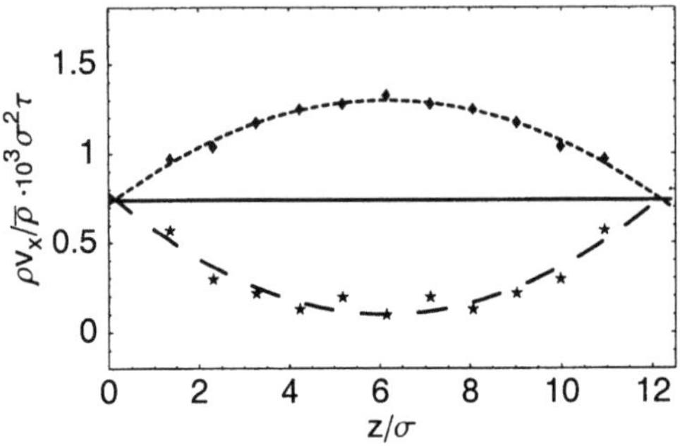

Fig. 20.5. The observed mass flow in the situation where $m_a = 1.25\,\mathrm{m}$ and $m_b = 0.75\,\mathrm{m}$. The lines are a parabolic fit to the data. The dashed line corresponds to a negative pressure gradient down the channel (along x) and the dotted line corresponds to a positive pressure gradient down the channel. Note that the lines intercept the left and right side of the channel at a finite positive velocity

molecules (equivalent to a body force in Navier–Stokes equation). Figure 20.5 shows the result. As can be seen, for a positive (negative) pressure gradient, one obtains a parabolic flow profile down (up) the channel typical of Poiseuille flow. The unusual feature is that these velocity profiles do *not* intercept the left and right side of the channel at zero velocity. In fact, even without a pressure gradient, one obtains a flat positive flow profile down the channel (solid line). Clearly, any attempt to fit these curves to a single slip length in the Navier slip condition (20.14) will fail. The finite mass transport in the absence of a pressure gradient is due to diffusion of the heavier molecules down the channel. This mass flux is well described by the relation $\rho v = -D\nabla\rho$ [17] and any full description of the boundary condition should include such terms.

An even more dramatic breakdown of the Navier condition comes in the case where one component preferentially wets the wall (in the case discussed above a and b have the same interaction with the walls). To examine this situation we use a system where the two particles are indistinguishable except for their labels. For particles a (b) we now change the wall-fluid interaction at the top (bottom) wall so that it is purely repulsive by truncating the potential at its minimum rather than at 2.2σ for particles b (a). Figure 20.6 shows the velocity produced by changing the wetting properties. One sees a remarkably strong shear flow (~m/s) with significant "slip" at the stationary walls. This slip has the wrong sign for the Navier condition and its magnitude is inconsistent with results for convective flows in the absence of concentration gradients.

The driving force for the flow in Fig. 20.6 comes from the externally imposed concentration gradient and the variation of the interfacial free energy of the walls with concentration. As can be seen from the figure, along the walls the system prefers the more strongly wetting fluid species. The differ-

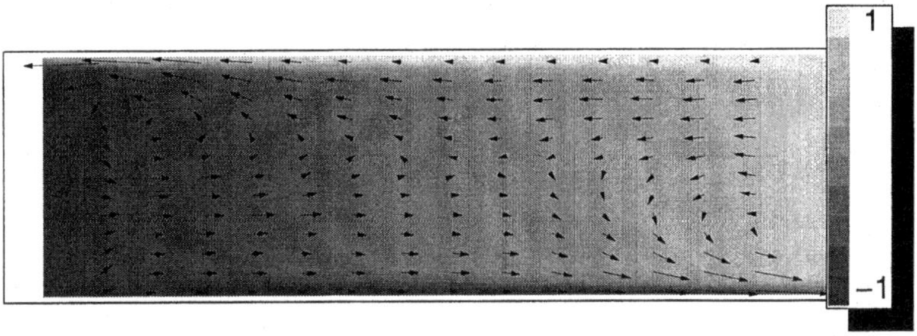

Fig. 20.6. Convective mass flow (vectors) and order parameter ϕ (shading) for a case where the wall at $z = 0$ preferentially wets a molecules and the other wall preferentially wets b

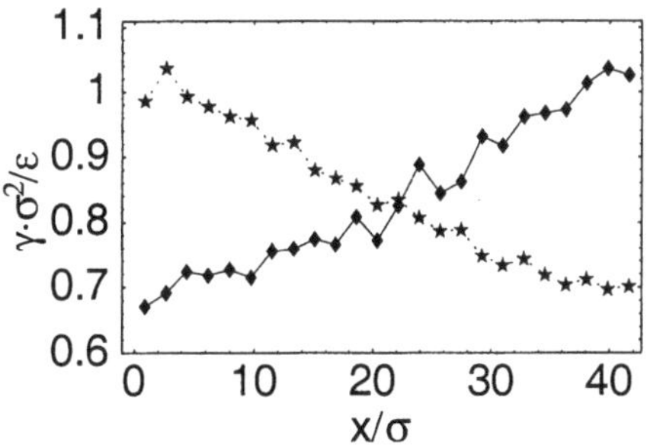

Fig. 20.7. Interfacial free energy along the top (diamonds) and bottom (stars) walls for the situation depicted in Fig. 20.6

ence between the concentration at the wall and in the center of the channel increases with x. This increase leads to a rise in interfacial tension. Values of the surface tension γ, calculated using the mechanical definition of Kirkwood and Buff [10], are shown in Fig. 20.7. The boundary condition relating the shear stress on the wall $\sigma_{xz}|_w$ to the viscous stress in the fluid $\sigma_{xz}|_f$ is analogous to that used for Marangoni flow at a two fluid boundary [24]:

$$\sigma_{xz}|_w - \sigma_{xz}|_f = \partial_x \gamma \,. \tag{20.15}$$

It can be shown [18] that there is also a velocity discontinuity given by an integral of the Marangoni stress over the interfacial region that produces the slip. This can be incorporated into a generalized Navier slip condition by adding a source term proportional to $\partial_x \gamma$. These boundary conditions determine the net stress on the wall. The result is consistent with stresses measured directly in our simulations, and ranges from 50 to 75% of $\partial_x \gamma$ (~MPa) for the cases we have studied. A detailed description and derivation of the boundary conditions and how they can be used to create a nanomotor or nanopump can be found in [18].

20.6 Conclusions

The approach we have outlined here of mapping molecular dynamics simulations of fluid–fluid interfaces onto mesoscale continuum theories for binary fluids is quite general and can be applied to more complex fluids. In order to fully model a complex fluid one must examine not only the interface profiles but also the microscopic stress, as we have done here. This allows a complete mapping from the length scales of molecular dynamics simulations onto a fully dynamic mesoscale model suitable for a lattice Boltzmann or other mesoscale simulation method. As we have seen, many typical assumptions of mesoscale models such as incompressibility fail at interfaces and this has important consequences for the resulting surface tension and interfacial dynamics.

Boundary conditions at solid-liquid interfaces are often taken for granted but can contain very unexpected terms when actually examined closely with a molecular dynamics simulation. Diffusive mass transport can contribute to a significant average velocity at the wall. Further, the presence of concentration gradients along the wall in the general case of non-neutral wetting can result in significant Marangoni-type forces which drive convective flow. These effects should be readily applicable to the design of new micro-fluidic devices and may be relevant to the function of numerous biological systems.

In the near future, the present work will be expanded to examine mesoscopic boundary conditions for immiscible binary fluids and other complex fluids such as liquid crystals.

Acknowledgement

This material is based upon work supported by the National Science Foundation under Grant No. 0083286. We thank Intel Corporation for donating the workstations used for our simulations, which were performed using LAMMPS from Sandia National Laboratories.

References

1. H.-P. Deutsch and K. Binder: Macromolecules **25**, 6214 (1992); and references therein
2. G.S. Grest, M.-D. Lacasse, K. Kremer, and A.M. Gupta: J. Chem. Phys. **105**, 10583 (1996)
3. M.-D. Lacasse, G.S. Grest, and A.J. Levine: Phys. Rev. Lett. **80**, 309 (1998). Although not explicitly noted, a clear systematic difference in the surface tension measured by capillary waves and through the mechanical definition is seen in Fig. 3
4. S. Barsky and M.O. Robbins: Phys. Rev. E **63**, 021801 (2001); ibid. **65**, 021808 (2002)

5. M.P. Allen and D.J. Tildesley: *Computer Simulation of Liquids* (Clarendon Press, Oxford 1987)
6. S.J. Plimpton, R. Pollock, M. Stevens: "Particle-Mesh Ewald and rRESPA for Parallel Molecular Dynamics Simulations", in *Proc. of the Eighth SIAM Conference on Parallel Processing for Scientific Computing, Minneapolis, MN, March 1997*; S.J. Plimpton: J. Comp. Phys. **117**, 1 (1995)
7. J.H. Irving and J.G. Kirkwood: J. Chem. Phys. **18**, 817 (1950)
8. L. Mistura: J. Chem. Phys. **83**, 3633 (1985); Inter. J. Thermophys. **8**, 397 (1987); J.S. Rowlinson: Proc. R. Soc. London Ser. A **402**, 67 (1985); J. Chem. Thermodyn. **25**, 449 (1993)
9. E. Wajnrjb, A.R. Altenberger, and J.S. Dahler: J. Chem. Phys. **103**, 9782 (1995)
10. J.G. Kirkwood and F.P. Buff: J. Chem. Phys. **17**, 338, (1949). This expression would need to be modified if our walls were not rigid
11. P.A. Thompson and M.O. Robbins: Phys. Rev. Lett. **63**, 766 (1989); P.A. Thompson, W.B. Brindkerhoff, and M.O. Robbins: J. Adhesion Sci. Techno. **7**, 535 (1993)
12. S. Toxvaerd and J. Stecki: J. Chem. Phys. **102**, 7163 (1995); ibid. **103**, 4352 (1995)
13. E. Diaz-Herrera, J. Alejandre, G. Ramirez-Santiago, and F. Forstmann: J. Chem. Phys. **110**, 8084 (1999)
14. G. Grest, M.-D. Lacasse, and M. Murat: in *Computer Simulation Studies in Condensed Matter Physics X*, D.P. Landau, S.P. Lewis, and H.B. Schüttler (Eds.) (Springer, Berlin Heidleberg New York 1998) p. 23
15. I. Napari, A. Laaksonen, V. Talanquer, D.W. Oxtoby: J. Chem. Phys. **110**, 5906 (1999)
16. C. Denniston and M.O. Robbins: *Mapping molecular models to continuum theories for partially miscible fluids*, submitted to Phys. Rev. E
17. C. Denniston and M.O. Robbins: Phys. Rev. Lett. **87**, 178302 (2001)
18. C. Denniston and M.O. Robbins: *Matching continuum boundary conditions to molecular dynamics simulations for a miscible binary fluid*, submitted to Phys. Rev. E
19. P.A. Thompson and M.O. Robbins: Phys. Rev. A **41**, 6830 (1990)
20. S.R. de Groot and P. Mazur: *Non-Equilibrium Thermodynamics*, (Dover Pub., New York 1984)
21. J.-L. Barrat and L. Bocquet: Phys. Rev. Lett. **82**, 4671 (1999); Faraday Discuss. **112**, 119 (1999)
22. J. Koplik and J.R. Banavar: Phys. Rev. Lett. **80**, 5125 (1998); ibid. **82**, 1334 (1999)
23. V. Ganesan and H. Brenner: Phys. Rev. Lett. **82**, 1333 (1999); Phys. Rev. E **61**, 6879 (2000)
24. L.D. Landau and E.M. Lifshitz: *Fluid Mechanics*, (Pergamon Press, Oxford 1987); S.H. Davis: Annu. Rev. Fluid Mech. **19**, 403 (1987); A.A. Golovin, A.A. Nepomnyashchy and L.M. Pismen: J. Fluid Mech. **347**, 317 (1997)

21 Electrokinetic Phenomena Revisited: A Lattice–Boltzmann Approach

J. Horbach

Institut für Physik, Johannes Gutenberg–Universität, Staudinger Weg 7, D–55099 Mainz, Germany

Abstract. The Lattice–Boltzmann method (LBM) is an efficient tool to solve the Navier–Stokes equations. Based on this method we have developed a scheme to investigate electrokinetic phenomena in charged colloidal suspensions. The equations of motion that are solved are the so-called electrokinetic equations, i.e. a set of partial differential equations that couple the gradient of the electrostatic potential to the hydrodynamic flow by means of a mean field theory. These equations have been extensively used to study electroviscous phenomena for the limit of a weakly charged sphere in an unbounded electrolyte. We demonstrate that our method can be applied beyond these limit. As an example we discuss the sedimentation of an array of charged spheres.

21.1 Introduction

In colloidal suspensions very different time and length scales are present due to the very different sizes of the solvent and the colloidal particles (for recent reviews on colloidal systems see [1–3]). Therefore, it is a very challenging task to study the dynamics of such systems by means of a computer simulation. For example, in a molecular dynamics computer simulation that accounts explicitly for the solvent particles one would spend most of the simulation time solving the equations of motion of the solvent particles since the movement of the colloidal particles is typically orders of magnitude slower than that of the solvent. But often one is not interested in the details of the solvent and one tries to circumvent the explicit simulation of it. This can be achieved, for instance, by describing the interactions between the colloidal particles by means of an effective potential. However, in this approach one completely neglects hydrodynamic interactions between the colloidal particles that stem from the momentum transport through the solvent. It has been shown in many experimental and theoretical studies that hydrodynamic interactions are crucial to understand transport phenomena in colloidal suspensions [4].

Hydrodynamic interactions could be taken into account if one describes the solvent on a coarse-grained continuum level. To this end, one may use the so-called Lattice–Boltzmann method (LBM) that has been developed as an efficient solver of the Navier–Stokes equations [5]. In the LBM a simple kinetic equation is solved on a lattice such that the Navier–Stokes equations are recovered. Recently, the LBM has been applied to compute dynamic properties of colloidal systems, such as the rotational and translational short-time dynamics [6–9], the diffusion of colloidal particles in confined geometry [10–12], and the dynamics of porous media [13–16]. Also other complex systems such as polymer solutions have been investigated by LBM [17,18].

Often the colloidal particles are charged and this affects strongly the static and dynamic properties. Thus, a realistic modelling of many colloidal systems requires the introduction of long-ranged Coulomb interactions. The simple restricted primitive model describes a system of macroions, counterions and coions and puts the degrees of freedom of the neutral part of the fluid into the dielectric constant. In a more realistic model one could describe the neutral part by a LB fluid. However, if one accounts for the charged species explicitly very big system sizes are required at least for highly charged macroions whereby the computation of the Coulomb interactions is in principle very time consuming due to their long range. E.g., if one considers a typical charge of each macroion of the order of 100–1000 e (e: charge of a proton) 100–1000 counterions of charge e for each macroion would be required and thus, a system of 1000 macroions would consist of at least 10^5 charged particles. Therefore, it would be very convenient to describe also the electrostatics on a continuum level and this is possible within a mean field description. To this end, one couples a time-dependent generalization of the Poisson–Boltzmann equation to the linearized Navier–Stokes equations. In this way one obtains the so-called electrokinetic equations of motion which are discussed in Sect. 21.2. These equations are the basis of our simulation method for charged colloidal systems which uses a LBM as a solver (Sect. 21.3). A LBM for charged species has already been proposed by Warren [19], but this method has not been worked out for colloidal systems. Although our LBM is different from that of Warren it was inspired by his approach (see [20]).

We have applied our method to study the sedimentation of an array of macroions in an electrolyte solution. The results for the sedimentation velocity of the latter system are shown in Sect. 21.4. Eventually a few conclusions are drawn in Sect. 21.5.

21.2 The Electrokinetic Equations of Motion

The so-called electrokinetic equations of motion are based on a purely phenomenological theory that combines the electrostatics on the level of the Poisson–Boltzmann equation with the linearized Navier–Stokes equations.

Consider a system of macroions with radius a in an electrolyte consisting of two ionic species that have charges $+z_1 e$ and $-z_2 e$, where z_1 and z_2 are the valencies of the ions. The densities of the ionic species, $\rho_1(\boldsymbol{r},t)$ and $\rho_2(\boldsymbol{r},t)$, are conserved quantities which can be expressed by the continuity equations

$$\frac{\partial \rho_s}{\partial t} = -\nabla \cdot \boldsymbol{J}_s \,, \quad s = 1, 2 \,. \tag{21.1}$$

For the currents $\boldsymbol{J}_s$ one makes the deterministic ansatz

$$\boldsymbol{J}_s = \rho_s \boldsymbol{u} - D_s \nabla \rho_s - z_s D_s \rho_s \nabla \hat{\Phi} \,, \tag{21.2}$$

where the first term describes a convection current (with the flow velocity $\boldsymbol{u}$) and the second and the third term are a diffusion current (D_s: diffusivity of the ions) and a current due to the electrostatic potential Φ (here in dimensionless form $\hat{\Phi} = \frac{e}{k_B T}\Phi$ with the Boltzmann constant k_B and the temperature T). Note that (21.2) is just a postulate which completely neglects statistical fluctuations and is thus a kind of mean-field description.

The electrostatic potential is further related to the ionic densities by the Poisson equation,

$$\nabla^2 \hat{\Phi} = -4\pi l_B \left(\sum_{s=1}^{2} z_s \rho_s + \sigma \right) , \tag{21.3}$$

where σ denotes the charge density of a macroion. The Bjerrum length $l_B = e^2/(4\pi\epsilon k_B T)$ (ϵ: dielectric constant) is the distance between two ions at which the Coulomb energy of one ion due to the other one is equal to $k_B T$.

If $\boldsymbol{u}$ is set to zero, (21.1) and (21.2) yield a Boltzmann distribution for the ionic densities in equilibrium ($\partial \rho_s / \partial t = 0$),

$$\rho_s(\boldsymbol{r}) = \overline{\rho}_s \exp\left(-z_s \hat{\Phi}(\boldsymbol{r}) \right) . \tag{21.4}$$

By putting the Boltzmann distribution, (21.4), into the Poisson equation (21.3) one obtains the Poisson–Boltzmann equation. Moreover, if the exponential function in (21.4) is linearized, (21.3) can be solved analytically (Debye–Hückel theory) and the result is a Yukawa potential,

$$\hat{\Phi}(\boldsymbol{r}) = K \frac{\exp(-\kappa r)}{r} , \tag{21.5}$$

where the Debye parameter κ is defined by

$$\kappa = \sqrt{4\pi l_B \sum_s z_s^2 \overline{\rho}_s} . \tag{21.6}$$

The potential (21.5) is often also a good approximation in the non-linear case but the prefactor K is then different from the one in the Debye–Hückel limit (see e.g. [4]).

We still have to specify an equation of motion for the total mass current of the fluid, $\rho \boldsymbol{u} \equiv (\sum_s \rho_s + \rho_n)\boldsymbol{u}$, where ρ_n denotes the density of the neutral part of the fluid. We assume in the following that $\rho_n \gg \rho_1, \rho_2$ which means that the total density of the fluid, ρ, is essentially distributed homogeneously everywhere in the fluid (i.e. also close to a macroion). Moreover, we consider a fluid that can be described by the linearized Navier–Stokes equations at low Reynolds numbers. Thus the equation for $\rho \boldsymbol{u}$ with a body force term due to the electrostatic potential is

$$\frac{\partial \rho \boldsymbol{u}}{\partial t} = \nu \nabla^2 \rho \boldsymbol{u} - \nabla p - k_B T \sum_s z_s \rho_s \nabla \hat{\Phi} , \tag{21.7}$$

where p is the pressure and ν the kinematic viscosity.

21.3 Lattice–Boltzmann Method

In this section we show how (21.1), (21.2), (21.3) and (21.7) can be solved numerically by means of a LBM. The central quantity in LBM is the one-particle distribution function $n_i(\boldsymbol{r},t)$ that decribes the number of particles on a lattice node $\boldsymbol{r}$ at time t with a discrete velocity $\boldsymbol{c}_i$. The discrete space of velocities $\{\boldsymbol{c}_i\}$ is chosen such that no artificial anisotropic terms appear in the hydrodynamic equations as derived from the kinetic equation for n_i. In our case the velocity space consists of 18 vectors of which, from a given lattice node, six point to the nearest and 12 to the next-nearest neighbors on a simple cubic lattice. It is one possible choice of a velocity space that exhibits the required isotropy [5].

The calculation of $n_i(\boldsymbol{r},t)$ consists of two steps, a collision and a propagation step. In the collision step the interaction between the particles is taken into account and this results in a post-collision function $n_i^\star(\boldsymbol{r},t^\star)$ at the collision time $t^\star$. In the propagation step n_i is updated by

$$n_i(\boldsymbol{r}+\boldsymbol{c}_i,t+1) = n_i^\star(\boldsymbol{r},t^\star)\,. \tag{21.8}$$

In this equation the lattice constant, the time step, and the mass of the particle is set to unity. The density $\rho(\boldsymbol{r},t)$ and the mass current $\boldsymbol{j} \equiv \rho\boldsymbol{u}$ are given by the zeroth and first moment of n_i, respectively, $\rho(\boldsymbol{r},t) = \sum_i n_i(\boldsymbol{r},t)$, $\boldsymbol{j}(\boldsymbol{r},t) = \sum_i n_i(\boldsymbol{r},t)\boldsymbol{c}_i$.

For the charged system a one-particle distribution function for each ionic species and a neutral part is required. The density of the neutral part is chosen such that its value at a given lattice point is much higher than that of the ionic densities to guaranty that the total density is essentially homogeneous through the fluid. We make the following ansatz for the post-collision function $n_i^{s\star}$ for the counter- and coions, $s = +,-$, respectively, and the neutral part, $s = n$ (we also take into account rest particles by the index $i = 0$):

$$n_i^{s\star}(\boldsymbol{r},t^\star) = \frac{\mathrm{w}_i\gamma_s}{24}\rho_s(\boldsymbol{r},t)\left(1+\frac{1}{c_{\mathrm{sv}}^2\rho'(\boldsymbol{r},t)}\boldsymbol{j}(\boldsymbol{r},t)\cdot\boldsymbol{c}_i\right)$$
$$\text{for} \quad i = 1,\dots,18 \tag{21.9}$$
$$n_0^{s\star}(\boldsymbol{r},t^\star) = (1-\gamma_s)\,\rho_s(\boldsymbol{r},t)\,. \tag{21.10}$$

The factor w_i is a weighting factor which is equal to 2 for the $\boldsymbol{c}_i$ in the direction of nearest neighbors and equal to 1 for the remaining $\boldsymbol{c}_i$. So it satisfies the normalization constraint $\sum_{i=1}^{18}\frac{\mathrm{w}_i}{24} = 1$. For the following we define also $\mathrm{w}_0 := 0$. With the parameter $0 < \gamma_s \leq 1$ the diffusivity D_s of the particles of type s can be varied. The latter quantity is given by [21]

$$D_s = \frac{c_{\mathrm{sv}}^2}{2}\gamma_s\,. \tag{21.11}$$

c_{sv} is the sound velocity which is $1/\sqrt{2}$ for our model [21]. The density ρ' is defined by $\rho' = \sum_s \gamma_s\rho_s$.

The propagation step for our charged system is

$$n_i^s(\boldsymbol{r}+\boldsymbol{c}_i,t+1) = n_i^{s\star}(\boldsymbol{r},t^\star)\,, \tag{21.12}$$

$$n_0^s(\boldsymbol{r},t+1) = n_0^{s\star}(\boldsymbol{r},t^\star)\,. \tag{21.13}$$

Different propagation rules have to be established at the surface of the macroions and at walls. Here we use the bounce back rules suggested by Ladd [21] which lead to no-slip boundary conditions. In this scheme one puts a sphere, which represents a macroion, onto the lattice whereby its surface cuts links between lattice nodes. The boundary nodes are defined halfway along these links and the population functions n_i^s which point to the direction of the boundary nodes are reflected back during the propagation step. In the case of moving boundaries there is a momentum transfer between the boundary nodes and the fluid.

The aforementioned way of mapping a sphere onto the lattice introduces fluid inside and outside the sphere. In our scheme we assign the charge of the macroions in that the inner fluid is an electrolyte with net charge Z. Charge neutrality requires that the total charge of the outer fluid equals the one of the inner fluid of the macroions. Of course, it is not allowed in our scheme that outer fluid leaks through the surface of the sphere. Therefore, only small movements of a macroion are possible such that the center of mass of the sphere can be fixed, and only a momentum transfer with the fluid takes place.

The densities ρ_s and the total mass current $\boldsymbol{j}$ cannot be inferred simply from the zeroth and first moments of the n_i^s's because we have to take into account their coupling to the gradient of the electrostatic potential. If ρ_s and $\boldsymbol{j}$ are calculated as follows,

$$\rho_s(\boldsymbol{r},t+1) = \sum_{i=0}^{18}\Big(n_i^s(\boldsymbol{r},t+1) - \frac{z_s}{2}\frac{\mathrm{w}_i\gamma_s}{24}\rho_s(\boldsymbol{r}-\boldsymbol{c}_i,t)\nabla\hat{\Phi}(\boldsymbol{r}-\boldsymbol{c}_i,t)\cdot\boldsymbol{c}_i\Big) \tag{21.14}$$

and

$$\boldsymbol{j}(\boldsymbol{r},t+1) = \sum_{s=1}^{3}\Big(\sum_{i=1}^{18} n_i^s(\boldsymbol{r},t+1)\boldsymbol{c}_i - c_{\mathrm{sv}}^2 z_s\gamma_s\rho_s(\boldsymbol{r},t+1)\nabla\hat{\Phi}(\boldsymbol{r},t+1)\Big)\,. \tag{21.15}$$

we are consistent with (21.1), (21.2) and (21.7) in the continuous limit. This can be explicitly shown by means of a Chapman–Enskog expansion (see [20]). Of course, $\nabla\hat{\Phi}$ does not couple to the neutral part of the fluid. This is guaranteed in (21.14) and (21.15) by setting $z_s = 0$ for $s = n$.

Finally we have to solve the Poisson equation at each time step in order to determine the electrostatic potential from the ionic densities ρ_+ and ρ_-.

For this purpose we use a successive over-relaxation (SOR) scheme in which one looks for the stationary solution of a diffusion equation [22] whereby one introduces an acceleration parameter $1 < \omega < 2$. The potential is then determined as the iterative solution of the following equation:

$$\hat{\Phi}_{h+1}(\boldsymbol{r},t) = \omega\left[\sum_i \left(\frac{\mathrm{w}_i}{24}\hat{\Phi}_h(\boldsymbol{r}-\boldsymbol{c}_i,t) + \pi l_\mathrm{B}\sum_{s=+,-} z_s\rho_s(\boldsymbol{r},t)\right)\right] + (1+\omega)\hat{\Phi}_h(\boldsymbol{r},t)\,. \tag{21.16}$$

Here h denotes the iteration time and the unit for an iteration step is set to unity. For a sufficient number of iteration steps $\hat{\Phi}_h$ converges to $\hat{\Phi}$. We have found that a suitable choice of ω is 1.45. The gradient of $\hat{\Phi}$ is calculated by

$$\nabla\hat{\Phi}(\boldsymbol{r}) = -\sum_i \frac{\mathrm{w}_i}{12}\hat{\Phi}(\boldsymbol{r}-\boldsymbol{c}_i)\boldsymbol{c}_i\,. \tag{21.17}$$

Note that the velocity space $\boldsymbol{c}_i$ of the lattice Boltzmann equations is also used to solve the Poisson equation for the electrostatic potential. More details on the calculation of the potential $\hat{\Phi}$ can be found in [20].

21.4 Sedimentation of an Array of Spheres

A sphere with radius a, that moves through a viscous fluid due to a gravitational field $\boldsymbol{g}$, experiences a drag force. The Stokes law relates this force to the velocity $\boldsymbol{U}_\mathrm{St}$ of the sphere [23]

$$\boldsymbol{F}_\mathrm{d} = \xi\boldsymbol{U}_\mathrm{St} \tag{21.18}$$

with

$$\xi = 6\pi\eta a\,. \tag{21.19}$$

ξ is the friction coefficient and η the shear viscosity of the fluid. The force $\boldsymbol{F}_\mathrm{d}$ is equal to the gravitational force on the sphere,

$$\boldsymbol{F}_\mathrm{g} = \frac{4}{3}\pi a^3(\rho_\mathrm{p} - \rho)\boldsymbol{g}\,, \tag{21.20}$$

where ρ_p and ρ denote the density of the sphere and the fluid, respectively. Therefore, from $\boldsymbol{F}_\mathrm{g} = \boldsymbol{F}_\mathrm{d}$ the sedimentation velocity follows as

$$\boldsymbol{U}_\mathrm{St} = \frac{2a^2(\rho_\mathrm{p} - \rho)}{9\eta}\boldsymbol{g}\,. \tag{21.21}$$

Equations (21.18) and (21.19) hold for a neutral spherical particle in an unbounded, incompressible fluid if one assumes no-slip boundary conditions at the surface of the sphere.

The situation is more complicated if the sphere is charged and moves through an electrolyte solution. Then the particle is surrounded by an electrical double layer that loses its spherical symmetry due to the fluid motion. This results in an electrical dipole field pointing in the direction opposite to the motion of the particle and the friction on the particle is increased as compared to (21.19) by additional contributions that depend on the properties of the electrolyte solution and the charge of the sphere Ze.

An analytical calculation of the sedimentation velocity of a weakly charged sphere in an unbounded electrolyte was first given by Booth [24]. Booth starts with the ansatz

$$\frac{U}{U_{\mathrm{St}}} = 1 + \sum_{k=1}^{\infty} c_k Z^k \,, \tag{21.22}$$

and considers only the term in the lowest non-vanishing order, Z^2,

$$\frac{U}{U_{\mathrm{St}}} = 1 + c_2 Z^2 \,. \tag{21.23}$$

The calculation of c_2 is then based on the electrokinetic equations (see Sect. 21.2) whereby the Poisson equation is solved in the Debye–Hückel limit. The final expression for c_2 is as follows [24,25]:

$$c_2 = -\frac{k_{\mathrm{B}} T l_{\mathrm{B}}}{72\pi a^2 \eta} \frac{\sum_s z_s^4 \overline{\rho}_s D_s^{-1}}{\sum_s z_s^2 \overline{\rho}_s} f(\kappa a) \,. \tag{21.24}$$

$f(\kappa a)$ can be found in the original publication [24] (see also [25]). It is a complicated function of exponential integrals of different order n, $E_n(x) = x^{n-1} \int_x^{\infty} dt\, t^{-n}\, \mathrm{e}^{-t}$.

Unfortunately, the sedimentation problem has not yet been solved analytically for the more general case of a system with a finite volume fraction of charged spheres beyond the limit that these spheres are only weakly charged. The next step in this context is to consider an array of charged spheres, a system which is well suited to be simulated by our LBM. To this end, we consider a macroion in a unit cell of length L with a fixed position. It represents one particle in an array of spheres on a simple cubic lattice interacting with its periodic images. The volume fraction of macroions is given by $\varphi = 4\pi a^3/(3L^3)$. In order to compute the sedimentation velocity one has to introduce a constant gravitational force in our LB equations. After a time of the order $\tau_s = L^2/\nu_{\mathrm{eff}}$ (ν_{eff}: effective, kinematic viscosity in the presence of the spheres) a steady state is reached and one can determine the average flow velocity in the unit cell. The latter is then divided by the average flow velocity of the corresponding neutral system to yield U/U_0, i.e. the ratio of the sedimentation velocity of the charged and the neutral system, respectively. It should be emphasized that U_0 is equal to the Stokes result, (21.21), only in the limit $\varphi \to 0$. For finite φ U_0 is the sedimentation velocity of an array of neutral spheres on a simple cubic lattice that is given for small φ by [26]

$$U_0 = U_{\mathrm{St}}(1 - 1.76\varphi^{1/3}) \,. \tag{21.25}$$

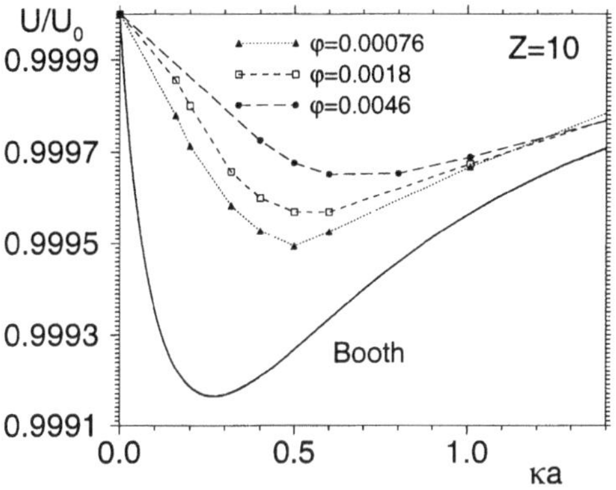

Fig. 21.1. U/U_0 for the volume fractions $\varphi = 0.0046$, 0.0018, and 0.00076 as a function of κa. The charge of the macroion is set to $Z = 10$. The solid line is the result from Booth's theory

We have determined U/U_0 as a function of κa for three different volume fractions, $\varphi = 0.00076$, 0.0018, and 0.0046. The radius of the macroions is fixed to $a = 4.5$. The diffusivities of the ionic species are set to $D_1 = 0.165$ for the counterions and $D_2 = 0.25$ for the coions. The ratio $D_1/D_2 = 0.66$ is equal to $D_{\mathrm{Na}}/D_{\mathrm{Cl}}$ in sodium chloride. The Bjerrum length is chosen to be $l_{\mathrm{B}} = 0.4$. In order to vary κa from 0.15 to 1.5 we change $\sum_s \overline{\rho}_s$ from 0.00025 to 0.02211. These densities are small as compared to that of the neutral fluid, $\rho_{\mathrm{n}} = 20$. Thus, the total density of the fluid is not changed significantly by varying κa.

Figure 21.1 shows U/U_0 as a function of κa for $Z = 10$ and the different φ as well as the Booth result which holds in the limit $\varphi \to 0$. It is rationalized below that $Z = 10$ is low enough for the approximation, (21.23), to hold. We infer from Fig. 21.1 that the relative reduction of the sedimentation velocity due to the charges is only of the order of 10^{-4} for $Z = 10$. U/U_0 exhibits a minimum around $\kappa a \approx 0.3$ that moves to slightly higher values with increasing φ. The occurrence of this minimum stems from the subtle interplay of the flow field surrounding the macroion and the electrostatic interactions that arise from the distortion of the Debye layer.

Figure 21.1 shows also that U/U_0 increases with increasing φ in the range of κa under consideration. This is remarkable since this behavior competes with the pure packing effect as described by (21.25) which predicts U to decrease by increasing φ. This means that with a suitable choice of the system parameters the electroviscous effect could win over the pure packing effect for small φ and also U and not only the reduced velocity U/U_0 could increase with φ. In this way one could understand the behavior of the electrophoretic mobility μ in an experiment of charged colloidal spheres in water that has been reported by Evers et al. [27]. These authors find indeed an increase of

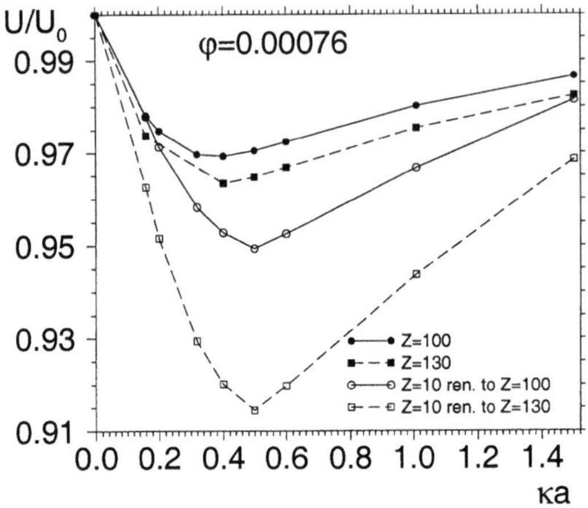

Fig. 21.2. U/U_0 at $\varphi = 0.00076$ as a function of κa for $Z = 100$ and $Z = 130$ (filled symbols). The open symbols show data for $Z = 10$ that are renormalized to $Z = 100$ and $Z = 130$ (see text)

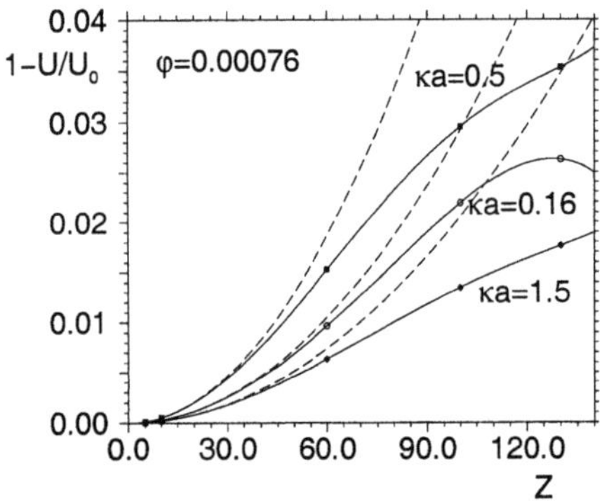

Fig. 21.3. $1 - U/U_0$ for the indicated values of κa. The solid lines are fites with functions $g(Z) = AZ^2 + BZ^4 + CZ^6$. The dashed lines show only the term $\propto Z^2$ of the fit functions

μ while the packing fraction of the colloidal particles increases from 10^{-6} to 2×10^{-3}.

Up to now we have only considered the case of a weakly charged macroion with $Z = 10$. But of course our method allows to go beyond this limit. So we can test up to which values of Z the approximation, (21.23), holds. If (21.23) would be valid, one could renormalize U/U_0 as a function of κa for a given charge $Z = Z_{\text{old}}$ to a new one $Z = Z_{\text{new}}$ by multiplying $(1 - U/U_0)$ by $Z^2_{\text{new}}/Z^2_{\text{old}}$. Figure 21.2 compares data for $\varphi = 0.00076$, that are renormalized from $Z = 10$ to $Z = 100$ and $Z = 130$, with the corresponding simulation results for the latter two values of Z. We see that the renormalized curves differ strongly from the "exact" ones in that the amplitude around the minimum is underestimated and the position of the minimum is at a slightly larger value in the renormalized curves.

To make more quantitative prdictions about the range of validity of Booth's theory we plot in Fig. 21.3 $(1 - U/U_0)$ as a function of Z for the indicated values of κa. The solid lines in this figure are fits of the form $g(Z) = c_2 Z^2 + c_4 Z^4 + c_6 Z^6$ where c_2, c_4, and c_6 are fit parameters. From the comparison of the different functions $g(Z)$ with only the leading term $\propto Z^2$ (dashed curve in Fig. 21.3) we can conclude that the corrections to Booth's theory become important for $Z > 50$.

21.5 Conclusions

We have developed a LBM for charged species with which it is possible to describe dynamical properties of charged colloidal particles in an electrolyte solution. In this method the so-called electrokinetic equations of motion are solved numerically that describe the hydrodynamics of the electrolyte on the level of a mean-field theory. These equations are the basis of many recent studies on electrokinetic phenomena in colloids. However, most of the latter investigations were restricted to very dilute systems and/or to weakly charged macroions. In contrast to that, our simulation method can go beyond these limits. This has been demonstrated in that we have applied our method to the determination of the sedimentation velocity of an array of charged spheres on a simple cubic lattice. We have discussed our results along Booth's theory for a weakly charged macroion in an unbounded electrolyte. Our LBM could be used to test theories that go beyond the Booth level. We mention that very recently also a mode coupling theory with hydrodynamic interactions was shown to be in agreement with Booth's theory [28]. This theory is also able to consider colloidal systems at finite volume fractions.

Acknowledgments

It is a pleasure to thank Daan Frenkel for a fruitful collaboration. Financial support by the Deutsche Forschungsgemeinschaft (Grant No. HO 2231/1-1) is gratefully acknowledged.

References

1. P.N. Pusey, in: *Liquids, Freezing and the Glass Transition.* Eds.: J.P. Hansen, D. Levesque and J. Zinn-Justin, Les Houches. Session LI, 1989 (North-Holland, Amsterdam 1991)
2. A.K. Arora and B.V.R. Tata: Adv. Coll. Int. Sci. **78**, 49 (1998)
3. H. Löwen: J. Phys.: Condens. Matter **13**, R415 (2001)
4. G. Nägele: Phys. Rep. **272**, 215 (1996)

5. S. Succi: *The Lattice Boltzmann Equation for Fluid Dynamics and Beyond* (Clarendon Press, Oxford 2001)
6. A.J.C. Ladd: Phys. Rev. Lett. **70**, 1339 (1993)
7. M.H.J. Hagen, D. Frenkel, and C.P. Lowe: J. Chem. Phys. **109**, 275 (1998)
8. M.H.J. Hagen, D. Frenkel, and C.P. Lowe: Physica A **272**, 376 (1999)
9. C.P. Lowe, D. Frenkel, and A.J. Masters: J. Chem. Phys. **103**, (1995)
10. M.H.J. Hagen, I. Pagonabarraga, C.P. Lowe, and D. Frenkel: Phys. Rev. Lett. **78**, 3785 (1997)
11. I. Pagonabarraga, M.H.J. Hagen, C.P. Lowe, and D. Frenkel: Phys. Rev. E **58**, 7288 (1998)
12. I. Pagonabarraga, M.H.J. Hagen, C.P. Lowe, and D. Frenkel: Phys. Rev. E **59**, 4458 (1999)
13. U. Oxaal, E.G. Flekkoy, and J. Feder: Phys. Rev. Lett. **72**, 3514 (1994)
14. A. Koponen, D. Kandhai, E. Hellen, M. Alava, A. Hoekstra, M. Kataja, K. Niskanen, P. Sloot, and J. Timonen: Phys. Rev. Lett. **80**, 716 (1998)
15. C.P. Lowe and D. Frenkel: Phys. Rev. Lett. **77**, 4552 (1996)
16. D.L. Koch, R.J. Hill, and A.S. Sangani: Phys. Fluids **10**, 3035 (1998)
17. P. Ahlrichs and B. Dünweg: J. Chem. Phys. **111**, 8225 (1999)
18. P. Ahlrichs, R. Everaers, and B. Dünweg: Phys. Rev. E **64**, 040501 (2001)
19. P.B. Warren: Int. J. Mod. Phys. C **8**, 889 (1997)
20. J. Horbach and D. Frenkel: Phys. Rev. E **64**, 061607 (2001)
21. A.J.C. Ladd: J. Fluid Mech. **271**, 285 (1994); **271**, 311 (1994)
22. W.H. Press, S.A. Teukolsky, W.T. Vetterling, and B.P. Flannery: *Numerical recipes.* (Cambridge University Press, Cambridge 1992)
23. G.K. Batchelor: *An introduction to fluid dynamics.* (Cambridge University Press, Cambridge 2000)
24. F. Booth: J. Chem. Phys. **22**, 1956 (1954)
25. H. Ohshima, T.W. Healy, L.R. White, and R.W. O'Brien: J. Chem. Soc., Faraday Trans. 2, **80**, 1299 (1984)
26. H. Hasimoto: J. Fluid Mech. **5**, 317 (1957)
27. M. Evers, N. Garbow, D. Hessinger, and T. Palberg: Phys. Rev. E **57**, 6774 (1998)
28. M. Kollmann and G. Nägele: Europhys. Lett. **52**, 474 (2000); M. Kollmann and G. Nägele: J. Chem. Phys. **113**, 7672 (2000)

Part V

Methods and Algorithms

22 Binary Tree Summation Monte Carlo Method for Potts Models

J.-S. Wang[1,2], O. Kozan[2], and R.H. Swendsen[2]

[1] Singapore-MIT Alliance and Department of Computational Science, National University of Singapore, Singapore 119260, Republic of Singapore
[2] Department of Physics, Carnegie Mellon University, Pittsburgh, PA 15213, USA

22.1 Introduction

Efforts to develop better and more efficient algorithms for Monte Carlo simulations have a long history, in which the Fortuin–Kasteleyn (FK) transformation [1] for Potts models has played a pivotal role. The most common use of this transformation has been to create algorithms in which clusters of spins in the Potts model are flipped simultaneously [2,3]. In this paper, we present a new algorithm using the FK representation in the spirit of recent work by Newman and Ziff [4] on the percolation problem. The method produces independent samples and sums up a large number of configurations for each sweep. The partition function and thermodynamic averages for all values of the temperature T can be computed from a single run.

Consider the Potts model with the Hamiltonian,

$$H(\sigma) = -J \sum_{\langle i,j \rangle} \delta_{\sigma_i,\sigma_j} , \tag{22.1}$$

where the summation is over nearest neighbor pair, $\sigma_i = 1, 2, \ldots, q$. The FK transformation allows us to write the partition function in the percolation representation as

$$Z = \sum_{\Gamma} p^b (1-p)^{M-b} q^{N_c(\Gamma)} , \tag{22.2}$$

where we sum over all configurations of bonds connecting nearest neighbor sites, $p = 1 - \exp\bigl(-J/(kT)\bigr)$, b is the number of bonds present, $M = dN$ is the maximum number of possible bonds, and $N_c(\Gamma)$ is the total number of clusters for a given configuration of bonds.

In order to evaluate thermodynamic averages in this representation, we carry out the summation over all configurations in two steps. First, we sum over the number of bonds b, and then for each value of b, we sum over all configurations consistent with that number of bonds. Thus we can write

$$Q(p) = Z^{-1} \sum_{b} p^b (1-p)^{M-b} c_b \, Q_b , \tag{22.3}$$

where $c_b = \sum_{\Gamma_b} q^{N_c(\Gamma_b)}$ and

$$Q_b = \frac{1}{c_b} \sum_{\Gamma_b} q^{N_c(\Gamma_b)} Q(\Gamma_b), \tag{22.4}$$

where Γ_b is a configuration with exactly b bonds. If we could compute c_b and Q_b for some observable $Q(\Gamma)$ for every b, then we could compute the function at any values of p or T.

The problem then reduces to the computation of the normalization constants c_b and the expectation values in an ensemble with probability distribution proportional to $q^{N_c(\Gamma)}$. For the special case of $q = 1$ (bond percolation), the values of c_b are the binomial coefficients and the configurations with fixed number of bonds are weighted uniformly. For general case, we describe two sampling methods. The first is very simple, and useful conceptually, but exponentially inefficient. The second turns out to be very efficient.

22.2 A Survival and Death Process

Starting with an empty lattice, one sweep consists of repeated application of the following steps until the process dies:

1. Pick an unoccupied neighbor pair at random for the next bond.
2. If inserting a bond
   (a) does not change the cluster number ($\Delta N_c = 0$), accept the configuration;
   (b) merge two clusters, so that the cluster number decreases by 1, ($\Delta N_c = -1$), accept the configuration with probability $1/q$, or reject the configuration and terminate the process (and begin the next sweep from an empty lattice).
3. Take statistics of the survival configurations (with equal weights).

The probability distribution of the configurations generated in this manner with b bonds is

$$P(\Gamma_b) = \frac{b!(M-b)!}{M!} q^{N_c(\Gamma_b) - N_c(\Gamma_0)} . \tag{22.5}$$

The survival probability after b bonds is then

$$S_b = \sum_{\Gamma_b} P(\Gamma_b) = \frac{b!(M-b)!}{M!} c_b \, q^{-N} , \tag{22.6}$$

where N is the total number of sites. As the survival probability decays exponentially with the number of bonds added, we expect very poor statistics for large values of b. However, the eventual survival probability, S_M, is equal to q^{-N+1}, which is nonzero.

22.3 Binary Tree Summation Method

To deal with the exponentially decreasing survival probability, we have developed the following method for summing multiple bond sequences in a single sweep. For a configuration Γ_b we count the number of type-0 bonds n_0 of unoccupied pairs of sites that are on the same cluster, and the number of type-1 bonds n_1 of unoccupied pairs of sites that would connect two different clusters. Clearly $n_0 + n_1 + b = M$. At each step, we pick a type-1 bond with equal probability from among all current type-1 bonds. Starting with an empty lattice and merging clusters at each step, we continue until all sites are members of the same cluster. Along the way, we collect statistics $Q(i)$ for each of the N configurations generated.

After a sweep has been carried out, we construct all possible paths that a full simulation of both type-0 and type-1 bonds would have taken if we had joined clusters in exactly the same sequence as in the actual simulation, but had also inserted a random number of type-0 bonds at each step. At each choice between types of bond, the type-1 bond is given a relative weight of n_1/q and a type-0 bond n_0. The total weight for a path starting from an empty lattice to a particular configuration is the product of the factors n_0 or n_1/q depending on the path taken.

To understand this algorithm, imagine that we had actually followed the branching process. At each step, a configuration may split into two configurations, one with a type-1 bond added (with probability $1/n_1$) and one with a type-0 bond added (with probability of $1/n_0$). The probability of appearance of a particular path is $\prod[n_{f(k)}(\Gamma_k)]^{-1}$, where $f(k) = 0$ or 1 depending on the choice of type-0 or 1 bond. In taking statistics, we have weighted with the inverse of the factor, multiplied by additional factors of q proportional to the number of clusters $N_c(\Gamma)$. The net effect is the required sample average with overall weight $q^{N_c(\Gamma)}$. Now the key observation is that we do not need an explicit simulation for the type-0 bonds, since the type-0 bonds have no effect on the measured quantities.

The binary tree summation algorithm has several attractive features. First, the usual slowing down due to correlation between samples is absent. Each sweep is independent. Second, for each sweep, data for a very large number of samples are collected. Although they are highly correlated, an exponentially large number of paths can be summed efficiently with $O(N^2)$ operations per sweep. Third, unlike multicanonical simulations [5] or the flat histogram method [6], there are no unknown weighting factors to determine.

22.4 Implementation

Let $w(b, i)$ be the weight of the total contributions from all possible paths to the state specified by the number of bonds b and merge sequence number i.

These quantities can be calculated from the starting condition $w(0,i)=\delta_{i,0}$, the constraint that $n_0(b,i)\geq 0$, and the recursive equation

$$w(b+1,i) = w(b,i)n_0(b,i) + w(b,i-1)n_1(b,i-1)/q\,, \tag{22.7}$$

where $n_1(b,i)=n_1(i)$, $n_0(b,i)=n_0(i)-b+i$. The value $w(b,i)$ is nonzero only for $b\geq i$. The computation of the weights $w(b,i)$ is similar to that of binomial coefficients. The final contribution to the statistics at b number of bonds is

$$Q_b = \langle W_b\rangle^{-1}\Big\langle \sum_{i=0}^{N-1} w(b,i)Q(i)\Big\rangle\,, \tag{22.8}$$

where the average is over simulation sweeps, and $W_b=\sum_i w(b,i)$ is the total weight at a given b. We can then show that

$$\langle W_b\rangle = \frac{M!}{(M-b)!}S_b\,, \tag{22.9}$$

which allows us to compute c_b. The result is numerically identical to computing the conditional survival probability S_{b+1}/S_b from the expectation value of $Q=(n_0+n_1/q)/(M-b)$.

During the simulation, the values of n_0 and n_1 can be updated efficiently. For each cluster, we keep a list of unoccupied bonds with other clusters. When two clusters (A and B) are joined, we merge the smaller one with the larger one, and remove and count the number of bonds n_{AB} connecting the clusters. We update according to $n_0 \leftarrow n_0 + n_{\mathrm{AB}} - 1$, $n_1 \leftarrow n_1 - n_{\mathrm{AB}}$. The timing of our program shows that this part of the algorithm scales nearly linearly with number of sites N, as expected.

22.5 Results and Discussions

We note that the coefficients c_b are related to the density of states $n(E)$, which gives the coefficients of the partition function polynomial in the variable $\exp\big(-J/(kT)\big)$. In the FK percolation representation, it becomes a polynomial in $p/(1-p)$. By a proper change of variables, we can find exact results [7] of c_b for the two-dimensional Ising model.

We define the following errors to test our method against exact results for the two-dimensional Ising model:

$$\epsilon_0 = \left|q^{N-1}c_M/c_0 - 1\right|, \tag{22.10}$$

$$\epsilon_1 = \frac{1}{M}\sum_{b=0}^{M}\left|c_b/c_b^{\mathrm{exact}} - 1\right|, \tag{22.11}$$

$$\epsilon_{\mathrm{MAX}}^{Q} = \max_T \left|Q(T) - Q^{\mathrm{exact}}(T)\right|, \tag{22.12}$$

$$\epsilon_{\mathrm{AVE}}^{Q} = \int_0^1 dx \left|Q\big(T(x)\big) - Q^{\mathrm{exact}}\big(T(x)\big)\right|, \quad T(x) = \frac{x}{1-x}\,. \tag{22.13}$$

Table 22.1. Errors for 10^6 sweeps with respect to the exact results of an $L \times L$ Ising model. While ϵ_1 is taken from an average over many runs, the other results are from a single run. The CPU times (on a 1.53 GHz Athlon) are in units of 10^{-6} second per sweep per site

L	4	8	16	32	50
cpu t	1.88	2.81	6.92	24.0	72.6
ϵ_1	0.0000634	0.000178	0.00049	0.0032	0.031
$\epsilon^E_{\mathrm{MAX}}$	0.000128	0.000113	0.00012	0.0015	0.0068
$\epsilon^E_{\mathrm{AVE}}$	0.0000184	0.0000103	0.0000055	0.000046	0.00014
$\epsilon^C_{\mathrm{MAX}}$	0.000306	0.00046	0.00124	0.0177	0.096
$\epsilon^C_{\mathrm{AVE}}$	0.000034	0.000031	0.000040	0.00050	0.0020

Due to a special cancellation for this algorithm, ϵ_0 is exactly zero. The error ϵ_1, along with the maximum and average errors in the energy and specific heat, are listed in Table 22.1. Since the algorithm asymptotically takes $O(N^2)$ operations per sweep, a fair comparison with other methods should compare the total CPU times. A comparable N-fold way transition matrix Monte Carlo (TMMC) run took 1.9 microsecond per sweep per site on the same machine. Thus, the present method is superior for small lattices with linear size $L \leq 16$. For $L = 32$ it is comparable to TMMC [6]. For much larger lattices, it becomes less favorable mainly due to the $O(N^2)$ nature of the algorithm. It is quite likely that we can speed up the computation using special properties of the weights.

Our method is applicable for Potts models with any number of states, including fractional or negative values. Work is currently in progress to apply this algorithm to a number of problems of interest.

References

1. P.W. Kasteleyn and C.M. Fortuin: J. Phys. Soc. Jpn Suppl. **26**, 11 (1969); C.M. Fortuin and P.W. Kasteleyn: Physica **57**, 536 (1972)
2. R.H. Swendsen and J.-S. Wang: Phys. Rev. Lett. **58**, 86 (1987)
3. U. Wolff: Phys. Rev. Lett. **62**, 361 (1989); Nucl. Phys. **B322**, 759 (1989)
4. M.E.J. Newman and R.M. Ziff: Phys. Rev. Lett. **85**, 4104 (2000); Phys. Rev. E **64**, 016706 (2001)
5. B.A. Berg and T. Neuhaus: Phys. Rev. Lett. **68**, 9 (1992)
6. J.-S. Wang and R.H. Swendsen: J. Stat. Phys. **106**, 245 (2002)
7. P.D. Beale: Phys. Rev. Lett. **76**, 78 (1996)

23 Probability-Changing Cluster Algorithm for Two-Dimensional Systems with the Vector Order Parameter

Y. Tomita and Y. Okabe

Department of Physics, Tokyo Metropolitan University,
Hachioji, Tokyo 192-0397, Japan

Abstract. The newly proposed probability-changing cluster (PCC) Monte Carlo algorithm is extended for the study of systems with the vector order parameter. We use Wolff's idea of the embedded cluster formalism for assigning clusters. The Kosterlitz–Thouless transitions for the two-dimensional XY and q-state clock models are studied systematically by using the PCC algorithm.

23.1 Introduction

In numerical studies, efficient algorithms are important for getting the necessary information. Recently we proposed an effective cluster algorithm, which is called the probability-changing cluster (PCC) algorithm, of tuning the critical point automatically [1]. It is an extension of the Swendsen–Wang algorithm [2], but we change the probability of cluster update depending on the observation whether clusters are percolating or not percolating.

The two-dimensional (2D) XY model shows a unique phase transition, the Kosterlitz–Thouless (KT) transition [3,4]. It does not have a true long-range order, but the correlation function decays as a power of the distance at all the temperatures below the KT transition point. José et al. [5] studied the effect of the q-fold symmetry-breaking fields on the 2D XY model; this is essentially the same as treating the q-state clock model, where only the discrete values are allowed for the angle θ_i of the XY spins. The 2D q-state clock model has two phase transitions of the KT type at T_1 and T_2 ($T_1 < T_2$) for $q > 4$. There is an intermediate XY-like phase between an ordered phase ($T < T_1$) and a disordered phase ($T > T_2$).

We may raise the following questions: (i) Can the PCC algorithm be used for the problem of the vector order parameter, such as the XY model? (ii) Can it be applied to the analysis of the transition of the KT type? (iii) Can it work even if the system shows two or more phase transitions?

The purpose of the present paper is to answer these questions. We extend the PCC algorithm so as to treat systems with the vector order parameter. We use the finite-size scaling (FSS) analysis based on the KT form of the correlation length. We investigate the KT transition of the 2D XY model and both phase transitions at T_1 and T_2 for the $q = 6, 8, 12$ clock models. The detailed report will be given in a separate paper [6].

23.2 PCC Algorithm and FSS Analysis

We treat the Hamiltonian given by

$$\mathcal{H} = -J \sum_{\langle i,j \rangle} \boldsymbol{S}_i \cdot \boldsymbol{S}_j \,, \tag{23.1}$$

where $\boldsymbol{S}_i$ is a planar unit vector, $(\cos\theta_i, \sin\theta_i)$, at site i; θ_i takes the value of $[0, 2\pi)$ for the XY model, and $2\pi p_i/q$ with $p_i = 0, 1, \cdots, q-1$ for the q-state clock model.

We use Wolff's idea of the embedded cluster formalism [7] to extend the PCC algorithm to systems with the vector order parameter. We project $\boldsymbol{S}_i$ onto a unit vector $\boldsymbol{e}_1$ and another unit vector $\boldsymbol{e}_2$, perpendicular to $\boldsymbol{e}_1$, as

$$\boldsymbol{S}_i = \boldsymbol{e}_1 \cos\phi_i + \boldsymbol{e}_2 \sin\phi_i \,, \tag{23.2}$$

where ϕ_i is the angle measured from the axis of the vector $\boldsymbol{e}_1$; a unit vector $\boldsymbol{e}_1$ is randomly chosen by every Monte Carlo step. Every site shares the common vectors, $\boldsymbol{e}_1$ and $\boldsymbol{e}_2$. Then, the Hamiltonian, (23.1), is rewritten as

$$\mathcal{H} = -\sum_{\langle i,j \rangle} (J_{ij}^{(1)} \epsilon_i^{(1)} \epsilon_j^{(1)} + J_{ij}^{(2)} \epsilon_i^{(2)} \epsilon_j^{(2)}) \tag{23.3}$$

with ϕ_i-dependent positive effective couplings $J_{ij}^{(1)}$ and $J_{ij}^{(2)}$ for two sets of Ising variables $\{\epsilon_i^{(1)}\}$ and $\{\epsilon_i^{(2)}\}$. Then, we can use the Kasteleyn–Fortuin (KF) cluster representation for the Ising spins [8]. To make the KF cluster, we connect the bonds of parallel Ising spins with the probability

$$p_{ij}^{(1,2)} = 1 - \exp(-2\beta J_{ij}^{(1,2)}) \,, \tag{23.4}$$

where $\beta = 1/k_B T$.

In the PCC algorithm, the cluster representation is used in two ways. First, we flip all the spins on any KF cluster to one of two states, that is, +1 or −1. Second, we change the KF probability, (23.4), depending on the observation whether clusters are percolating or not. It is based on the fact that the spin–spin correlation function $G(\boldsymbol{r}_i - \boldsymbol{r}_j)$ becomes nonzero for $|\boldsymbol{r}_i - \boldsymbol{r}_j| \to \infty$ at the same point as the percolation threshold. For the XY model in the embedded cluster formalism, $G(\boldsymbol{r}_i - \boldsymbol{r}_j)$ is written as

$$G(\boldsymbol{r}_i - \boldsymbol{r}_j) = \langle \boldsymbol{S}_i \cdot \boldsymbol{S}_j \rangle = \left\langle A_{ij}\, \Theta^{(1)}(\boldsymbol{r}_i, \boldsymbol{r}_j) \right\rangle + \left\langle B_{ij}\, \Theta^{(2)}(\boldsymbol{r}_i, \boldsymbol{r}_j) \right\rangle \,, \tag{23.5}$$

where $\langle \cdots \rangle$ represent the thermal average. The function $\Theta(\boldsymbol{r}_i, \boldsymbol{r}_j)$ is equal to 1 (0) if the sites i and j belong to the same (different) cluster, and A_{ij} and B_{ij} are some constants. Thus, the system is regarded as percolating if $\epsilon^{(1)}$ *or* $\epsilon^{(2)}$ Ising spins are percolating.

We may formulate the checking process of percolation in another way. Instead of choosing the vector $\boldsymbol{e}_1$ randomly in (23.2), choosing $\boldsymbol{e}_1$ as

$$\boldsymbol{M} = |\boldsymbol{M}| \, (\boldsymbol{e}_1 \cos\phi + \boldsymbol{e}_2 \sin\phi) \tag{23.6}$$

yields a new condition for the percolation, where $\boldsymbol{M} = \sum_i \boldsymbol{S}_i$, and ϕ is some *fixed* angle between the ordered direction $\boldsymbol{M}$ and $\boldsymbol{e}_1$. Then, $G(\boldsymbol{r}_i - \boldsymbol{r}_j)$ is expressed by the sum of two functions depending on ϕ; that is,

$$G(\boldsymbol{r}_i - \boldsymbol{r}_j) = g(\phi; \boldsymbol{r}_i - \boldsymbol{r}_j) + g(\pi/2 - \phi; \boldsymbol{r}_i - \boldsymbol{r}_j) . \tag{23.7}$$

Using (23.7), we check whether the system is percolating or not. This modification of the checking process is quite useful for the study of the low-temperature transition T_1 for the clock models.

Let us consider the size dependence of the transition temperature. We use the FSS analysis based on the KT form of the correlation length, $\xi \propto \exp(c/\sqrt{t})$ with $t = (T - T_{\mathrm{KT}})/T_{\mathrm{KT}}$. Using the PCC algorithm, we locate the temperature $T_{\mathrm{KT}}(L) = 1/k_B\beta_c(L)$ that the existence probability E_p is $1/2$. Then, using the FSS form of E_p, that is, $E_p = E_p(\xi/L)$, we have the relation

$$T_{\mathrm{KT}}(L) = T_{\mathrm{KT}} + \frac{c^2 T_{\mathrm{KT}}}{(\ln bL)^2} . \tag{23.8}$$

Next consider the magnetization $\langle m^2(L) \rangle$ at $T_{\mathrm{KT}}(L)$ to discuss the critical exponent η. We expect the FSS of the form $\langle m^2(L) \rangle \propto L^{-\eta}$, but there are small multiplicative logarithmic corrections [4,9], that is,

$$\langle m^2(L) \rangle = AL^{-\eta} (\ln b'L)^{-2r} . \tag{23.9}$$

We use (23.9) for the estimate of the exponent η.

23.3 Results

We have made simulations for the XY model on the square lattice with the system sizes L = 8, 16, 32, 64, 128, 256, and 512. We have also treated the q = 6, 8, 12 clock models. After 20,000 Monte Carlo sweeps of determining $\beta_c(L)$ we have made 10,000 Monte Carlo sweeps to take the thermal average; we have made 100 independent runs for each size.

In Fig. 23.1a, we plot $T_1(L)$ and $T_2(L)$ as a function of l^{-2} with $l = \ln bL$ for the $q = 6, 8, 12$ clock models. We represent the temperature in units of J/k_B. The error bars are smaller than the size of marks. Although we do not show the plot for the XY model here, the size dependence is almost the same as that for $T_2(L)$ of the clock models. The estimates of the transition temperatures including T_{KT} for the XY model based on (23.8) are tabulated in Table 23.1. Our estimate of T_{KT} of the XY model is 0.8933(6); the number in the parentheses denotes the uncertainty in the last digits. This value is consistent with the estimates of recent studies; 0.89213(10) by the Monte Carlo simulation [10], and 0.894 by the short-time dynamics [11]. The estimate of T_2 for the 6-state clock model is 0.9008(6), which is more precise than the previous estimates; 0.92(1) [12] and 0.90 [13]. Our estimate of T_1 for the 6-state clock model, 0.7014(11), is more precise again than the previous estimates; 0.68(2) [12] and 0.75 [13].

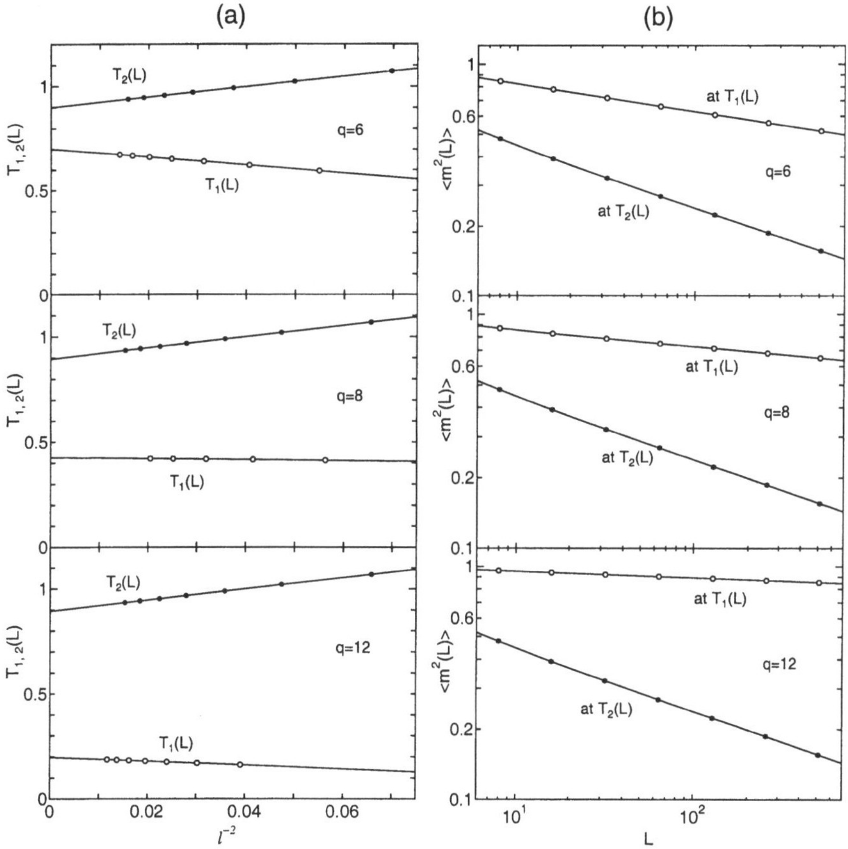

Fig. 23.1. (**a**) Plots of $T_1(L)$ and $T_2(L)$ of the 2D q-state clock models for $L = 8$, 16, 32, 64, 128, 256, and 512, where $l = \ln bL$. (**b**) Plots of $\langle m^2(L)\rangle$ at $T_1(L)$ and $T_2(L)$

In Fig. 23.1b, we give the plot of $\langle m^2(L)\rangle$ at $T_1(L)$ and $T_2(L)$ as a function of L in logarithmic scale. Using the FSS form with the multiplicative logarithmic corrections, (23.9), we estimate the exponent η and the logarithmic-correction exponent r. We tabulate the estimates of η and r including those for the XY model in Table 23.1. The estimates for the XY model are $\eta = 0.243(4)$ and $r = 0.038(5)$. This value of η is a little bit smaller than the theoretical prediction, $1/4$ ($= 0.25$). Our logarithmic-correction exponent r is compatible with Janke's result $r = 0.0560(17)$ [9], but different from the theoretical prediction $r = -1/16$ [4]. The estimate of η at T_2 for the 6-state clock model is 0.243(4), and that for r is 0.037(5). The estimate of η at T_1 for the 6-state clock model is 0.113(3), and it is more precise than the previous estimates of η; 0.100 [12] and 0.15 [13].

Table 23.1. Transition temperatures, exponents η, and correction exponents r for the 2D XY and q-state clock models

	T_2	$\eta(T_2)$	$r(T_2)$	T_1	$\eta(T_1)$	$r(T_1)$
$q = 6$	0.9008(6)	0.243(4)	0.037(5)	0.7014(11)	0.113(3)	0.017(4)
$q = 8$	0.8936(7)	0.243(4)	0.041(5)	0.4259(4)	0.0657(2)	0.0056(4)
$q = 12$	0.8937(7)	0.246(5)	0.034(6)	0.1978(5)	0.0270(5)	0.0026(7)
XY ($q = \infty$)	0.8933(6)	0.243(4)	0.038(5)	———	———	———

Let us discuss the q-dependence of transition temperatures and exponents. The transition temperature T_1 becomes lower with larger q; in the lowest order we find that $T_1 \propto 1/q^2$, which is consistent with the theoretical prediction [5]. The exponent η at T_2 is a universal constant, and compatible with the theoretical prediction $\eta = 1/4$. The estimates of η at T_1 remarkably coincide with the theoretical prediction $\eta = 4/q^2$; $1/9 = 0.111$ for $q = 6$, $1/16 = 0.0625$ for $q = 8$, and $1/36 = 0.0278$ for $q = 12$. This is the *first* systematic report of confirming the theoretical prediction.

23.4 Discussions

In summarizing, we have successfully applied the PCC algorithm [1] to the XY and clock models. We finally make a brief comment on the generalization of the PCC algorithm. The PCC algorithm can be applied to the quantum Monte Carlo simulation. The study of the quantum XY model using the generalized scheme of the PCC algorithm is now in progress.

Acknowledgments

We thank N. Kawashima, H. Otsuka, M. Itakura, and Y. Ozeki for valuable discussions. Special thanks are due to D.P. Landau for helpful discussions.

References

1. Y. Tomita and Y. Okabe: Phys. Rev. Lett. **86**, 572 (2001); J. Phys. Soc. Jpn. **71**, 1570 (2002)
2. R.H. Swendsen and J.S. Wang: Phys. Rev. Lett. **58**, 86 (1987)
3. J.M. Kosterlitz and D. Thouless: J. Phys. C **6**, 1181 (1973)
4. J.M. Kosterlitz: J. Phys. C **7**, 1046 (1974)
5. J.V. José, L.P. Kadanoff, S. Kirkpatrick, and D.R. Nelson: Phys. Rev. B **16**, 1217 (1977)
6. Y. Tomita and Y. Okabe: Phys. Rev. B **66**, 180401 (2002)
7. U. Wolff: Phys. Rev. Lett. **62**, 361 (1989)
8. P.W. Kasteleyn and C.M. Fortuin: J. Phys. Soc. Jpn. Suppl. **26**, 11 (1969); C.M. Fortuin and P.W. Kasteleyn: Physica **57**, 536 (1972)

9. W. Janke: Phys. Rev. B **55**, 3580 (1997)
10. P. Olsson: Phys. Rev. B **52**, 4526 (1995)
11. B. Zheng, M. Schulz, and S. Trimper: Phys. Rev. E **59**, R1351 (1999)
12. M.S.S. Challa and D.P. Landau: Phys. Rev. B **33**, 437 (1986)
13. A. Yamagata and I. Ono: J. Phys. A **24**, 265 (1991)

24 Flat Histogram Method of Wang–Landau and N-Fold Way

B.J. Schulz, K. Binder, and M. Müller

Johannes Gutenberg Universität, Staudinger Weg 7, D55099 Mainz, Germany

Abstract. We present a method for estimating the density of states of a classical statistical model. The algorithm successfully combines the Wang–Landau flat histogram method with the N-fold way in order to improve the efficiency of the original single-spin-flip version. We compare our data with the detailed numerical tests presented in a study by Wang and Swendsen where the original Wang–Landau method was tested against various other methods, especially the transition matrix Monte Carlo method (TMMC).

24.1 Introduction

While Monte Carlo methods in statistical thermodynamics already find broad application [1–3], the standard approach using the Metropolis algorithm [4] suffers from the drawback that the entropy of the simulated model system is not an output of the calculation. Furthermore, the convergence of the method in practice is often problematic due to the fact that systems with a rugged landscape of the (coarse-grained) free energy may get trapped in one valley of such a free energy landscape for a long time.

In order to overcome these difficulties, many interesting approaches (e.g., umbrella sampling [5], multicanonical Monte Carlo[6], expanded ensemble methods [7,8]) have been proposed in the literature, and some of these techniques emphasize the idea of directly sampling the energy density of states (e.g., [9–15]). Recently, F. Wang and Landau [12] proposed a Monte Carlo algorithm for classical statistical models which uses a random walk in energy space in order to obtain an accurate estimate for the density of states $g(E)$. This method is based upon the fact that a flat energy histogram $H(E)$ is produced if the probability for the transition to a state of energy E is proportional to $1/g(E)$. Initially, $g(E)$ is set equal to one for all energies. A spin is then chosen at random and flipped with probability $\min(1, g(E)/g(E'))$ whereby E' is the energy of the system with the chosen spin being overturned. The density of states $g(E)$ is not constant during the random walk, but is updated according to $g(E) \to g(E) \cdot f$ after each spin flip trial whether the spin is flipped or not. A histogram H(E) records how often a state of energy E is visited. In the beginning of the random walk the modification factor f can be as large as $e \simeq 2.7182818$.[1]. Each time the energy histogram

[1] If one chooses to sample $s(E) = \log_{10} g(E)$ the modification factor f becomes an increment $s(E) \to s(E) + \Delta s$ with $\Delta s = \log_{10} f \leq \log_{10} e \simeq 0.4342944$

satisfies the flatness criterion $H(E) \geq \epsilon \cdot \langle H(E) \rangle$, f is reduced according to $f \to f^{\frac{1}{2}}$ and $H(E)$ is reset to zero for all energies. The simulation is ended if f is close enough to one. To speed up simulations it is possible to perform several random walks on adjacent energy intervals on independent processors. Disadvantages of the single-spin-flip version of this algorithm are the small acceptance rates for energy intervals which contain the groundstate and low-lying excited states and the relatively large errors of $g(E)$ at right edges [20] of energy intervals as reported in [12,13]. Since this flat histogram method produces only a relative density of states $g(E)$, one has to normalize $g(E)$ in order to get the absolute density of states $\hat{g}(E)$. This can be done by using known values of the density of states, for example, the groundstate degeneracy or other constraints on $\hat{g}(E)$. J.S. Wang and Swendsen presented an efficient transition matrix Monte Carlo method (TMMC) [15] using the N-fold way which allows for the computation of the density of states with help of an optimization procedure. In particular, they found their method to be superior to the single-spin-flip Wang–Landau flat histogram method. Adopting the idea of Wang and Landau in the context of transition matrix resulted in an improvement of efficiency only for small system sizes up to $L = 8$. Larger systems however, showed the problem of sticking to a Gaussian distribution for the histogram. In the next section we will discuss how the algorithm of Wang and Landau can be successfully combined with the rejection-free N-fold way [13].

24.2 N-Fold Way

In the N-fold way [16] a flip occurs at each step of the algorithm and one then calculates the life-time of the resulting state. Consequently, observables become life-time weighted averages over the generated states.

Since the density of states $\hat{g}(E)$ can be very large especially for large system sizes, we consider $s(E) = \log_{10} g(E)$ (one can also use the natural ln) during simulations. In the beginning $s(E)$ is set to zero for all E. Initially the system may be in the state σ with energy $E \in I = [E_{min}, E_{max}]$, whereby I denotes the energy-range for which one wants to estimate $g(E)$. One then partitions all spins into classes according to the energy difference ΔE_i a spin flip will cause. For a two-dimensional nearest neighbor Ising model each spin belongs to one of only $M = 10$ classes. The total probability P of any spin of class i being overturned is given by

$$P(\Delta E_i) = n(\sigma, \Delta E_i) p(E \to E + \Delta E_i), \quad i = 1, ..., M\,, \tag{24.1}$$

whereby $n(\sigma, \Delta E_i)$ is the number of spins of state σ belonging to class i and for $p(E \to E + \Delta E_i)$ one has

$$p(E \to E + \Delta E_i) = \begin{cases} \min(1, g(E)/g(E + \Delta E_i)) & \text{if} \quad E + \Delta E_i \in I \\ 0 & \text{if} \quad E + \Delta E_i \notin I\,. \end{cases} \tag{24.2}$$

With help of the integrated probabilities for a spin flip within the first m classes

$$Q_m = \sum_{i \leq m} P(\Delta E_i)\,, \quad m = 1, ..., M \quad \text{and} \quad Q_0 = 0\,, \tag{24.3}$$

one selects a class by generating a random number $0 < r < Q_M$ and the spin to be overturned is then chosen from class m if $Q_{m-1} < r < Q_m$ with equal probabilities. Due to the flip, the spin and its interacting neighbors will change classes and correspondingly the numbers $n(\sigma, \Delta E_i)$ will differ from their predecessors. Finally, one has to determine the average lifetime τ of a state. The probability that the first random number would produce a flip is $\hat{P} = Q_M/N$, N being the number of spins. Therefore one has $\bar{P}_n = \hat{P}(1-\hat{P})^{n-1}$ for the probability $\bar{P}_n$ that exactly n random numbers will result in a new configuration, which leads to

$$\tau = \sum_{n=1}^{\infty} n\bar{P}_n = \sum_{n=1}^{\infty} n\hat{P}(1-\hat{P})^{n-1} = \frac{N}{Q_M}\,. \tag{24.4}$$

Now, we can state the steps which are carried out by the N-fold way version of the Wang–Landau method:

1. Choose an initial state and set $H(E) = 0$, $s(E) = 0$ for all E and $\Delta s_0 = \log_{10} e \simeq 0.4342944$ and also fix Δs_{final}.
2. Determine (update) the probabilities $p(E \to E + \Delta E_i)$ and the Q_m's of the (initial) configuration using (24.1), (24.2) and (24.3).
3. Calculate average life-time τ of (initial) state via (24.4).
4. Increment histogram, density of states and update Δs_i:

$$\begin{aligned} H(E) &\to H(E) + \tau \\ s(E) &\to s(E) + \Delta\bar{s} \\ \Delta s_i &\to \Delta s_{i+1} \end{aligned}$$

   with

$$\Delta\bar{s} = \begin{cases} \Delta s_i \cdot \tau & \text{if} \quad \Delta s_i \cdot \tau \leq \log_{10} e \simeq 0.4342944 \\ \log_{10} e & \text{if} \quad \Delta s_i \cdot \tau > \log_{10} e\,, \end{cases} \tag{24.5}$$

$$\Delta s_{i+1} = \begin{cases} \Delta s_i & \text{if} \quad \Delta s_i \cdot \tau \leq \log_{10} c \\ \Delta\bar{s}/\tau & \text{if} \quad \Delta s_i \cdot \tau > \log_{10} e\,, \end{cases} \tag{24.6}$$

   in case of $\Delta s_i/\Delta s_{i+1} > 2$ we set $\Delta s_{i+1} \to \Delta s_i/2$.
5. After some fixed number of sweeps check $H(E)$ and refine Δs_j according to $\Delta s_{j+1} = \Delta s_j/2$ if $H(E)$ is flat. Stop if histogram is flat for a $\Delta s_j \leq \Delta s_{final}$.
6. Look up the Q_m's ($p(E \to E + \Delta E_i)$ is not updated here).
7. Choose and flip spin as described above and go to 2.

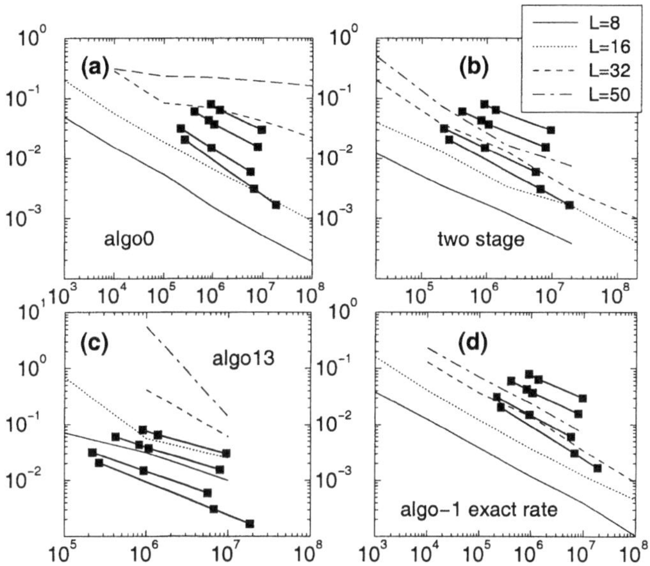

Fig. 24.1. Average error $\bar{\varepsilon}$ in the density of states $\hat{g}(E)$ vs. the number of sweeps (*thick solid lines* correspond to $L = 50, 32, 16, 8$ from top to bottom) in comparison to the data of [15] (see legend on right top). (**a**) TMMC with original flat histogram flip rates. (**b**) Algorithm *algo0* followed by *algo-1* which uses the estimated density of states from the first stage for the multi-canonical flip rates. (**c**) Implementation of the Wang–Landau flat histogram method (*algo13*) using single spin flips. (**d**) TMMC with exact multi-canonical flip rates

Equations (24.5) and (24.6) ensure that the increment $\Delta\bar{s}$ is always smaller or equal to $\log_{10} e$. Larger upper limits for $\Delta\bar{s}$ result in large statistical errors in $\hat{g}(E)$ [12]. Furthermore they can lead to $Q_M = 0$ in the context of the N-fold way and the convergence of the iteration procedure is lost. We have observed that adjusting Δs_i according to (24.5) and (24.6) only occurs in the first stage of iteration. Once the histogram is flat for the first time the increment $\Delta\bar{s}$ is simply $\Delta s_i \cdot \tau$. Controlling Δs_i in other ways may also work.

24.3 Tests and Conclusion

We have tested our algorithm with the two-dimensional nearest neighbor Ising model (square lattices up to $L = 50$) and also a more complex thin Ising film (e.g. [17–19]) of size $32 \times 32 \times 6$ with antiparallel surface fields. The relative error $\bar{\varepsilon} = (1/[N-2]) \sum |\hat{g}(E)/\hat{g}_{\text{exact}}(E) - 1|$ in the density of states for the two-dimensional Ising model is depicted in Fig. 24.1. Our simulations show that the present algorithm is competitive in efficiency with the TMMC approaches proposed in [15] and that errors in the density of states at right edges of an energy interval are not systematically enhanced as in the case of the single-spin-flip version [12,13]. For a more detailed description of these simulations we refer the reader to [13].

Acknowledgement

This work received financial support from the Deutsche Forschungsgemeinschaft (DFG) under grant No. Bi314/16-2 and the Bundesministerium für Bildung und Forschung (BMBF) under grant No. 03N6015. We are grateful to David P. Landau for numerous stimulating discussions, and for providing preprints of [12] prior to publication. We are grateful to Jian-Sheng Wang and Robert H. Swendsen for sending us a preprint of [15].

References

1. K. Binder and D.W. Heermann: *Monte Carlo Simulations in Statistical Physics. An Introduction* (3rd edn. (Springer, Berlin Heidelberg New York 1997)
2. D. Frenkel and B. Smit: *Understanding Molecular Simulation: From Algorithms to Applications* (Academic Press, London 1996)
3. D.P. Landau and K. Binder: *A Guide to Monte Carlo Simulations in Statistical Physics* (Cambridge University Press, Cambridge 2000)
4. N. Metropolis, A.W. Rosenbluth, A.H. Teller, and E. Teller: J. Chem. Phys. **21**, 1032 (1953)
5. G.M. Torrie and J.P. Valleau: Chem. Phys. Lett. **28**, 578 (1974); G.M. Torrie and J.P. Valleau: J. Comp. Phys. **23**, 187 (1977)
6. B.A. Berg and T. Neuhaus; Phys. Lett. B **276**, 249 (1991); Phys. Rev. Lett. **68**, 9 (1992)
7. A.P. Lyubartsev, A.A. Martsinovsky, S.V. Sherkunov, and P.N. Vorontsov-Velyaminov: J. Chem. Phys. **96**, 1776 (1992)
8. Y. Iba: Int. J. Mod. Phys. C **12**, 623 (2001)
9. J. Lee: Phys. Rev. Lett. **71**, 211 (1993)
10. P.M.C. de Oliveira, T.J.P. Penna and H.J. Herrmann: Eur. Phys. J. B **1**, 205 (1998); P.M.C. de Oliveira, T.J.P. Penna, and H.J. Herrmann: Braz. J. Phys. **26**, 677 (1996); P.M.C. de Oliveira: Eur. Phys. J. B **6**, 111 (1998)
11. J.-S. Wang: Physica A **281**, 147 (2000)
12. F. Wang and D.P. Landau: Phys. Rev. Lett. **86**, 2050 (2001); Phys. Rev. E **64**, 056101 (2001)
13. B.J. Schulz, K. Binder, and M. Müller: Int. J. Mod. Phys. C **13**, 477 (2002)
14. J.-S. Wang: cond-mat/0103318
15. J.-S. Wang and R.H. Swendsen: J. Stat. Phys. **106**, 245 (2002)
16. A.B. Bortz, M.H. Kalos and J.L. Lebowitz: J. Comput. Phys. **17**, 10 (1975)
17. K. Binder, D.P. Landau, and A.M. Ferrenberg: Phys. Rev. Lett. **74**, 298 (1995); Phys. Rev. E **51**, 2823 (1995)
18. K. Binder, R. Evans, D.P. Landau, and A.M. Ferrenberg: Phys. Rev. E **53**, 5023 (1996)
19. A.M. Ferrenberg, D.P. Landau, and K. Binder: Phys. Rev. E **58**, 3353 (1998)
20. Such systematic errors can be avoided by sampling the density of states adequately at edges; B.J. Schulz, K. Binder and M. Müller: in preparation

25 Solving the Time-Dependent Maxwell Equations by Unconditionally Stable Algorithms

J.S. Kole, M.T. Figge, and H. De Raedt

Centre for Theoretical Physics and Materials Science Centre
University of Groningen, Nijenborgh 4,
NL-9747 AG Groningen, The Netherlands
E-mail: j.s.kole@phys.rug.nl, m.t.figge@phys.rug.nl, deraedt@phys.rug.nl
http://www.compphys.rug.nl

Abstract. We present a family of unconditionally stable algorithms, based on the Suzuki product-formula approach, that solve the time-dependent Maxwell equations in systems with spatially varying permittivity and permeability. Salient features of these algorithms are discussed. As an illustration we compute, as a function of cluster size, the spectrum of electromagnetic modes in a cluster of photonic bandgap material.

25.1 Introduction

The Maxwell equations describe the evolution of electromagnetic (EM) fields in space and time [1]. They apply to a wide range of different physical situations and play an important role in a large number of engineering applications. In many cases, numerical methods are required to solve Maxwell's equations, either in the frequency or time domain. For the time domain, a well-known class of algorithms is based on a method proposed by Yee [2] and is called the finite-difference time-domain (FDTD) method. These algorithms owe their popularity mainly due to their flexibility and speed while at the same time they are easy to implement. A limitation of Yee-based FDTD techniques is that their stability is conditional, depending on the mesh size of the spatial discretization and the time step of the time integration [1].

In this paper we describe a family of unconditionally stable algorithms that solve the time-dependent Maxwell equations (TDME) [3]. The key to the construction of these algorithms is the observation that orthogonal approximations to the (orthogonal) time evolution operator automatically yield unconditionally stable algorithms. The Lie–Trotter–Suzuki product formulae [5–7] provide the mathematical framework to construct orthogonal approximations to the time-evolution operator of the Maxwell equations.

25.2 Theory

We consider EM fields in a three-dimensional medium with spatially varying permittivity and/or permeability, surrounded by a perfectly conducting box. In the absence of free charges and currents, the EM fields in such a system satisfy Maxwell's equations [1]

$$\frac{\partial}{\partial t}\boldsymbol{B} = -\nabla\times\boldsymbol{E} \quad \text{and} \quad \frac{\partial}{\partial t}\boldsymbol{D} = \nabla\times\boldsymbol{H}\,, \tag{25.1}$$

$$\operatorname{div}\boldsymbol{B} = 0 \quad \text{and} \quad \operatorname{div}\boldsymbol{D} = 0\,, \tag{25.2}$$

where $\boldsymbol{H} = (H_x(\boldsymbol{r},t), H_y(\boldsymbol{r},t), H_z(\boldsymbol{r},t))^T = \mu\boldsymbol{B}$ denotes the magnetic field, $\mu = \mu(\boldsymbol{r})$ is the permeability, $\boldsymbol{E} = (E_x(\boldsymbol{r},t), E_y(\boldsymbol{r},t), E_z(\boldsymbol{r},t))^T = \boldsymbol{D}/\varepsilon$ the electric field, and $\varepsilon = \varepsilon(\boldsymbol{r})$ denotes the permittivity. For simplicity of notation, we will omit the spatial dependence on $\boldsymbol{r} = (x,y,z)^T$ unless this leads to ambiguities. On the surface of the perfectly conducting box the EM fields satisfy the boundary conditions $\mathbf{n}\times\boldsymbol{E} = 0$ and $\mathbf{n}\cdot\boldsymbol{H} = 0$, with $\mathbf{n}$ denoting the vector normal to a boundary of the surface [1].

Some important symmetries of the Maxwell equations (25.1)–(25.2) can be made explicit by introducing the fields $\boldsymbol{X}(t) = \sqrt{\mu}\boldsymbol{H}(t)$ and $\boldsymbol{Y}(t) = \sqrt{\varepsilon}\boldsymbol{E}(t)$. In terms of the fields $\boldsymbol{X}(t)$ and $\boldsymbol{Y}(t)$, the TDME read

$$\frac{\partial}{\partial t}\begin{pmatrix}\boldsymbol{X}(t)\\ \boldsymbol{Y}(t)\end{pmatrix} = \begin{pmatrix}0 & -\frac{1}{\sqrt{\mu}}\nabla\times\frac{1}{\sqrt{\varepsilon}}\\ \frac{1}{\sqrt{\varepsilon}}\nabla\times\frac{1}{\sqrt{\mu}} & 0\end{pmatrix}\begin{pmatrix}\boldsymbol{X}(t)\\ \boldsymbol{Y}(t)\end{pmatrix} \equiv \mathcal{H}\begin{pmatrix}\boldsymbol{X}(t)\\ \boldsymbol{Y}(t)\end{pmatrix}. \tag{25.3}$$

Writing $\Psi(t) = (\boldsymbol{X}(t), \boldsymbol{Y}(t))^T$, (25.3) becomes $\frac{\partial}{\partial t}\Psi(t) = \mathcal{H}\Psi(t)$ It is easy to show that $\mathcal{H}$ is skew-symmetric, i.e. $\mathcal{H}^T = -\mathcal{H}$, with respect to the inner product $\langle\Psi|\Psi'\rangle \equiv \int_V \Psi^T\cdot\Psi'\,d\boldsymbol{r}$, where V denotes the volume of the enclosing box. The formal solution of the TDME is given by $\Psi(t) = U(t)\Psi(0) = e^{t\mathcal{H}}\Psi(0)$, where $\Psi(0)$ represents the initial state of the EM fields. The operator $U(t) = e^{t\mathcal{H}}$ determines the time evolution. By construction $\|\Psi(t)\|^2 = \langle\Psi(t)|\Psi(t)\rangle = \int_V \left[\varepsilon\boldsymbol{E}^2(t) + \mu\boldsymbol{H}^2(t)\right] d\boldsymbol{r}$, relating the length of $\Psi(t)$ to the energy density $w(t) \equiv \varepsilon\boldsymbol{E}^2(t) + \mu\boldsymbol{H}^2(t)$ of the EM fields [1]. As $U(t)^T = U(-t) = U^{-1}(t) = e^{-t\mathcal{H}}$ it follows that $\langle U(t)\Psi(0)|U(t)\Psi(0)\rangle = \langle\Psi(t)|\Psi(t)\rangle = \langle\Psi(0)|\Psi(0)\rangle$. Hence the time-evolution operator $U(t)$ is an orthogonal transformation, rotating the vector $\Psi(t)$ without changing its length $\|\Psi\|$. In physical terms this means that the energy density of the EM fields does not change with time, as expected on physical grounds [1].

25.3 Unconditionally Stable Algorithms

A numerical procedure that solves the TDME necessarily starts by discretizing the spatial derivatives. This maps the continuum problem described by $\mathcal{H}$

onto a lattice problem defined by a matrix H. Ideally, this mapping should not change the basic symmetries of the original problem. The underlying symmetry of the TDME suggests to use matrices H that are real and skew-symmetric. Adopting the staggered Yee grid and central differences for the spatial derivatives [2,1], the resulting matrix H is indeed skew-symmetric and the EM fields satisfy the required boundary conditions [3]. As the discretization procedure is not essential for what follows, we omit these technicalities here and refer the reader to [3].

Formally the time evolution of the EM fields on the lattice is given by $\Psi(t+\tau) = U(\tau)\Psi(t) = e^{\tau H}\Psi(t)$. In practice, a numerical procedure solves the TDME by making use of an approximation $\tilde{U}(t)$ to the true time evolution $U(t)$. A necessary and sufficient condition for an algorithm to be unconditionally stable is that $\|\tilde{U}(t)\Psi(0)\| \leq \|\Psi(0)\|$. In other words, the length of $\Psi(t)$ should be bounded, for arbitrary initial condition $\Psi(t=0)$ and for any time t [8]. By choosing for $\Psi(0)$ the eigenvector of $\tilde{U}(t)$ that corresponds to the largest eigenvalue of $\tilde{U}(t)$, it follows that the algorithm will be unconditionally stable by construction if and only if the largest eigenvalue of $\tilde{U}(t)$ (denoted by $\|\tilde{U}(t)\|$) is less or equal than one [8]. If the approximation $\tilde{U}(t)$ is itself an orthogonal transformation, then $\|\tilde{U}(t)\| = 1$ and the numerical scheme will be unconditionally stable.

A systematic approach to construct orthogonal approximations to matrix exponentials is to make use of the Lie–Trotter–Suzuki formula [5,6] and generalizations thereof [7,9]. For instance if $H = \sum_{i=1}^{p} H_i$

$$U_1(\tau) = e^{\tau H_1} \ldots e^{\tau H_p}\,, \tag{25.4}$$

might be a good approximation to $U(\tau)$ if τ is sufficiently small and, most importantly, if all the H_i are real and skew-symmetric, $U_1(\tau)$ is orthogonal by construction. Therefore, by construction, a numerical scheme based on (25.4) will be unconditionally stable. The Taylor series of $U(\tau)$ and $U_1(\tau)$ are identical up to first order in τ. The product-formula approach provides simple, systematic procedures to improve the accuracy of the approximation to $U(\tau)$ without changing its fundamental symmetries. For example the orthogonal matrix

$$U_2(\tau) = U_1(-\tau/2)^T U_1(\tau/2) = e^{\tau H_p/2} \ldots e^{\tau H_1/2} e^{\tau H_1/2} \ldots e^{\tau H_p/2}\,, \tag{25.5}$$

is correct up to second-order in τ [7,9]. Suzuki's fractal decomposition approach [7] gives a general method to construct higher-order approximations based on $U_1(\tau)$ or $U_2(\tau)$. A particularly useful approximation, correct up to fourth-order in τ, is given by [7]

$$U_4(\tau) = U_2(a\tau)U_2(a\tau)U_2((1-4a)\tau)U_2(a\tau)U_2(a\tau)\,, \tag{25.6}$$

where $a = 1/(4-4^{1/3})$. The approximations (25.4), (25.5), and (25.6) have proven to be very useful in many applications [6,9,13,11–18] and turn out to be equally useful for solving the TDME. In practice, an efficient implementation of the first-order scheme is all that is needed to construct the higher-order algorithms (25.5) and (25.6).

25.4 Simulation Results

Numerically the existence of photonic bandgaps is most easily studied by adopting periodic boundary conditions. However, in nature periodic boundary conditions are very hard to realize. Therefore it is of interest to study the dependence of photonic bandgaps as a function of system size with physically realizable boundary conditions. The method we outline above is well-suited for this purpose. It does not rely on periodic boundary conditions nor on the use of the Fourier transform of the unit cell. We consider three-dimensional clusters of touching spheres filled with air in a dielectric background. The centers of the spheres form an FCC lattice, see Fig. 25.1. The procedure we use to compute the (Local) Density of States (L)DOS is described in detail in [19].

First we compute the DOS for the system with periodic boundary conditions for a reference system containing $8 \times 8 \times 8$ FCC unit cells of linear size λ. The FCC lattice constant is set at $a = 3\lambda$ and, combined with a mesh size $\delta = 0.1\lambda$, ensures that the spheres of air are discretized with sufficient detail. The permittivity of the dielectricum $\epsilon = 11.9\epsilon_0$, where ϵ_0 is the dielectric constant of vacuum. The DOS of this system clearly exhibits one photonic bandgap, see Fig. 25.1 (dashed line) at $\omega a/2\pi c = 0.8$ where c is the speed of light in vacuum. Then we compute the LDOS of the FCC cluster of n spheres filled with air (without periodic boundary conditions) by using as an initial state, a random wave localized within the central sphere. In Fig. 25.1 we also show the results for $n = 19, 249, 1409$. It is clear that the finite-sized clusters also exhibit a photonic bandgap and that the strength of the bandgap strongly depends on n (note the logarithmic scale of the y-axis). Roughly speaking the minimum of the LDOS approaches its limiting value as $e^{-\alpha\sqrt{n}}$ where α is a material specific constant.

25.5 Conclusion

We have described a new family of algorithms to solve the time-dependent Maxwell equations. Salient features of these algorithms are [3,4]:

- rigorously provable unconditional stability for 1D, 2D and 3D systems with spatially varying permittivity and permeability,
- the order of accuracy in time and space can be systematically increased without affecting the unconditional stability,
- the exact conservation of the energy density of the electromagnetic field,
- easy to implement in practice.

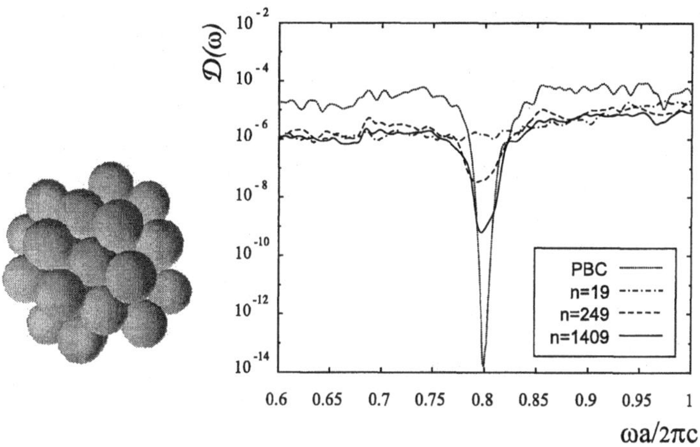

Fig. 25.1. Left: The 3D system: group of connected spheres of air (grey) in a dielectric background (white). The centers of the spheres form an FCC structure Right: Local density of states as function of frequency for 3D systems consisting of n spheres. PBC denotes the reference system with periodic boundary conditions

In view of the generic character of the approach discussed in this paper, it can be used to construct unconditionally stable algorithms that solve the equations for e.g. sound, seismic and elastic waves as well.

Acknowledgements

This work is partially supported by the Dutch "Stichting Nationale Computer Faciliteiten" (NCF).

References

1. M. Born and E. Wolf: *Principles of Optics* (Pergamon, Oxford 1964)
2. K.S. Yee: IEEE Transactions on Antennas and Propagation **14**, 302 (1966)
3. A. Taflove and S.C. Hagness: *Computational Electrodynamics – The Finite-Difference Time-Domain Method* (Artech House, Boston 2000)
4. J.S. Kole, M.T. Figge and H. De Raedt: Phys. Rev. E **64**, 066705 (2001)
5. H.F. Trotter: Proc. Am. Math. Soc. **10**, 545 (1959)
6. M. Suzuki, S. Miyashita, and A. Kuroda: Prog. Theor. Phys. **58**, 1377 (1977)
7. M. Suzuki: J. Math. Phys. **26**, 601 (1985); *ibid.* **32** 400 (1991)
8. G.D. Smith: *Numerical solution of partial differential equations* (Clarendon, Oxford 1985)
9. H. De Raedt and B. De Raedt: Phys. Rev. A **28**, 3575 (1983)
10. H. De Raedt: Comp. Phys. Rep. **7**, 1 (1987)
11. H. Kobayashi, N. Hatano, and M. Suzuki: Physica A **211**, 234 (1994)
12. H. De Raedt, K. Michielsen: Comp. in Phys. **8**, 600 (1994)
13. A. Rouhi, J. Wright: Comp. in Phys. **9**, 554 (1995)

14. B.A. Shadwick and W.F. Buell: Phys. Rev. Lett. **79**, 5189 (1997)
15. M. Krech, A. Bunker, and D.P. Landau: Comp. Phys. Comm. **111**, 1 (1998)
16. P. Tran: Phys. Rev. E **58**, 8049 (1998)
17. K. Michielsen, H. De Raedt, J. Przeslawski, and N. Garcia: Phys. Rep. **304**, 89 (1998)
18. H. De Raedt, A.H. Hams, K. Michielsen, and K. De Raedt: Comp. Phys. Comm. **132**, 1 (2000)
19. A. Hams and H. De Raedt: Phys. Rev. E **62**, 4365 (2000)
20. J.S. Kole, M.T. Figge and H. De Raedt: submitted to Phys. Rev. E

26 Chebyshev Method to Solve the Time-Dependent Maxwell Equations

H. De Raedt, K. Michielsen, J.S. Kole, and M.T. Figge

Institute for Theoretical Physics and Materials Science Centre
University of Groningen, Nijenborgh 4,
NL-9747 AG Groningen, The Netherlands
E-mail: deraedt@phys.rug.nl, kristel@phys.rug.nl,
j.s.kole@phys.rug.nl, m.t.figge@phys.rug.nl
http://www.compphys.rug.nl/
Version: 20 March 2002

Abstract. We present a one-step algorithm to solve the time-dependent Maxwell equations for systems with spatially varying permittivity and permeability. We compare the results of this algorithm with those obtained from unconditionally stable algorithms and demonstrate that for a range of applications the one-step algorithm may be orders of magnitude more efficient than multiple time-step, finite-difference time-domain algorithms. We discuss both the virtues and limitations of this one-step approach.

26.1 Introduction

Many applications in physics and engineering require numerical methods to solve the time-dependent Maxwell equations (TDME) [1]. A popular approach is the finite-difference time-domain (FDTD) method [1] based on a proposal by Yee [2]. It is flexible, fast and easy to implement. A limitation of Yee-based FDTD techniques is that their stability is conditional, depending on the mesh size of the spatial discretization and the time step of the time integration [1].

Recently we have introduced a family of unconditionally stable algorithms to solve the TDME [3–5]. The operator that governs the time evolution of the electromagnetic (EM) fields is orthogonal and can be written as the matrix exponential of a skew-symmetric matrix [3,5]. Orthogonal approximations to the time-evolution operator yield unconditionally stable algorithms by construction [13]. Details of the construction of such algorithms can be found elsewhere [3,5].

A limitation of both the Yee-based [1] and our unconditionally stable algorithms [3,5] is that the amount of computational work required to propagate the EM fields for long times may be prohibitive for a class of important applications, such a bioelectromagnetics and VLSI design [1,6,7]. The basic reason for this is that in order to maintain a reasonable degree of accuracy during the time integration, the time step has to be relatively small. In addition, in the Yee-based approach the time step is also limited by the Courant number [1].

A well-known alternative to time-stepping is to use Chebyshev polynomials to construct approximations to the time-evolution operator [8–10,12]. In this paper we make use of these rapidly converging polynomial approximations to construct a one-step algorithm that solves the TDME. Also in this case the orthogonality of the time-evolution operator plays a key role in the construction of the algorithm.

26.2 One-step Slgorithm

In the absence of charges and currents, the TDME can be written as [3,5]

$$\frac{\partial}{\partial t}\Psi(t) = \mathcal{H}\Psi(t)\,, \tag{26.1}$$

where $\Psi(t)$ is a vector that contains all information on the magnetic fields $\boldsymbol{H} = (H_x(\boldsymbol{r},t), H_y(\boldsymbol{r},t), H_z(\boldsymbol{r},t))^T$ and electric fields $\boldsymbol{E} = (E_x(\boldsymbol{r},t), E_y(\boldsymbol{r},t), E_z(\boldsymbol{r},t))^T$. The operator $\mathcal{H}$ is a compact representation for the curl operators that appear in the Maxwell equations [3]. Its precise form is not important for what follows. We will only use the fact that $\mathcal{H}$ is skew-symmetric, i.e. $\mathcal{H}^T = -\mathcal{H}$, with respect to the inner product $\langle\Psi|\Psi'\rangle \equiv \int_V \Psi^T \cdot \Psi'\, d\boldsymbol{r}$, where V denotes the volume of the enclosing, perfectly conducting box. The formal solution of (26.1) is given by $\Psi(t) = U(t)\Psi(0) = e^{t\mathcal{H}}\Psi(0)$ where $\Psi(0)$ represents the initial state of the EM fields.

A numerical procedure that solves the TDME involves a discretization of the spatial derivatives. This maps the continuum problem described by $\mathcal{H}$ onto a lattice problem defined by a (sparse) matrix H. The underlying symmetry of the TDME suggests to use matrices H that are real and skew-symmetric. The discretization procedure itself is not essential for what follows as long as H is skew symmetric. We omit these technicalities here and refer the reader to [3].

The time evolution of the EM fields on the lattice is given by $\Psi(t+\tau) = U(\tau)\Psi(t) = e^{\tau H}\Psi(t)$. The final step in the construction of the numerical procedure to solve the Maxwell equation is to choose a (good) approximation of the time-evolution operator $U(\tau)$. In the case of the Chebyshev approach we proceed as follows.

First we "normalize" the matrix H. The eigenvalues of the skew-symmetric matrix H are pure imaginary numbers. Hence the eigenvalues of the Hermitian matrix $A = -iH$ are real and if a is one of these eigenvalues so is $-a$. The eigenvalues of A lie in the interval $[-\|A\|_2, \|A\|_2]$ where $\|A\|_2$ is the largest (in absolute value) eigenvalue of A [14]. Obviously $\|A\|_2$ is hard to find. However for our purposes we only need an upperbound to $\|A\|_2$. Since A is sparse it is easy to compute $\|A\|_1 \equiv \max_j \sum_i |A_{i,j}|$ and the upperbound follows from $\|A\|_2 \leq \|A\|_1$ [14]. Note that $\|A\|_1 = \|H\|_1$. By construction the eigenvalues of $B \equiv A/\|A\|_1 = -iH/\|H\|_1$ all lie in the interval $[-1,1]$. The time-evolution operator then reads $U(t) = e^{tH} = e^{izB}$ where $z = t\|A\|_1$.

Expanding the initial value $\Psi(0)$ in the (unknown) eigenvectors $\boldsymbol{b}_j$ of B we have $\Psi(t) = U(t)\Psi(0) = e^{izB}\Psi(0) = \sum_j e^{izb_j}\boldsymbol{b}_j\langle\boldsymbol{b}_j|\Psi(0)\rangle$, where the b_j denote the (unknown) eigenvalues of B. Using $e^{izx} = J_0(z) + 2\sum_{k=1}^{\infty} i^k \times J_k(z)T_k(x)$, where $J_n(z)$ is the Bessel function of integer order n and $T_n(x) = \cos(n \arccos x)$ is the n-th Chebyshev polynomial of the first kind [15], we obtain

$$e^{izB}\Psi(0) = \left[J_0(z)I + 2\sum_{k=1}^{\infty} J_k(z)\widetilde{T}_k(B)\right]\Psi(0). \tag{26.2}$$

Here I is the identity matrix and $\widetilde{T}_k(B) = i^k T_k(B)$ is a matrix-valued modified Chebyshev polynomial that is defined by the recursion

$$\widetilde{T}_0(B)\Psi(0) = \Psi(0)\,, \quad \widetilde{T}_1(B)\Psi(0) = iB\Psi(0), \tag{26.3}$$

$$\widetilde{T}_{k+1}(B)\Psi(0) = 2iB\widetilde{T}_k(B)\Psi(0) + \widetilde{T}_{k-1}(B)\Psi(0) \quad \text{for} \quad k \geq 1\,. \tag{26.4}$$

From (26.3) and (26.4) it is clear that $\widetilde{T}_k(B)\Psi(0)$ is real valued, as it should be in the case of the Maxwell equations.

In practice we will have to truncate the sum in (26.2), i.e. we will use only the first $K+1$ contributions to approximate $U(t)\Psi(0)$:

$$\Psi(t) = e^{tH}\Psi(0) \approx \left[J_0(t\|H\|_1)I + 2\sum_{k=1}^{K} J_k(t\|H\|_1)\widetilde{T}_k(B)\right]\Psi(0)\,. \tag{26.5}$$

As $\|T_k(B)\|_1 \leq 1$ by construction and $|J_k(z)| \leq |z|^k/2^k k!$ for z real [15], we may expect that the error of this approximation vanishes (exponentially) fast with increasing K if K is sufficiently large.

According to (26.5), performing one time step amounts to repeatedly using recursion (26.4) to obtain $\widetilde{T}_k(B)\Psi(0)$ for $k = 2,\ldots,K$, to multiply the elements of this vector by $J_k(t\|H\|_1)$ and to add all contributions. This procedure requires storage for two vectors of the same length as $\Psi(0)$ and some code to multiply such a vector by the sparse matrix H. The coefficients $J_k(t\|H\|_1)$ should be calculated to sufficiently high precision. The number K is fixed by requiring that $|J_k(t\|H\|_1)| < \kappa$ for all $k > K$. Here κ is a control parameter that determines the accuracy of the solution. Keeping κ fixed, it is evident that the larger $t\|H\|_1$, the larger K will have to be in order to keep the accuracy the same.

26.3 Simulation Results

In our simulations we measure distances in units of the wavelength λ and time in units of λ/c where c is the speed of light in vacuum. In Fig. 26.1

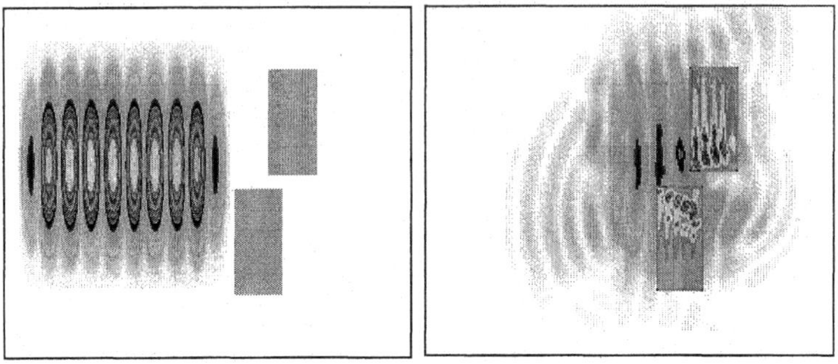

Fig. 26.1. Initial (left) and final distribution of the EM field intensity at $t = 5$ of the TM mode, as obtained from the numerical solution of the Maxwell equations by the one-step algorithm (26.5) with $K = 194$. The fourth-order unconditionally stable algorithm $T4S2$ [3,4] yields the same pictures (not shown)

Table 26.1. Performance of the one-step algorithm that solves the TDME for three different times t as obtained from simulations of the two-dimensional system depicted in Fig. 26.1. The fourth-order unconditionally stable algorithm $T4S2$ [3,4] with a time step τ was used to compute $\Psi(t,\tau)$. The vector $\Psi(t)$, obtained by the Chebyshev method, was taken as reference for the calculation of the error $e = \|\Psi(t) - \Psi(t,\tau)\|$. N_H is the number of times the operation $H\Psi$ was carried out

	$t=2$		$t=5$		$t=20$	
τ	e	N_H	e	N_H	e	N_H
0.100	2.5×10^{-3}	200	5.3×10^{-3}	500	1.8×10^{-2}	2000
0.010	4.0×10^{-7}	2000	8.5×10^{-7}	5000	2.9×10^{-6}	20000
0.001	4.0×10^{-11}	20000	8.5×10^{-11}	50000	2.9×10^{-10}	200000
Chebyshev	0	96	0	194	0	649

we present numerical results for a two-dimensional system of size 12×10, with mesh size $\delta = 0.1$. The number of lattice sites in the x and y-direction is $L_x = 239$ and $L_y = 199$ respectively. The shape of the initial wave (see Fig. 26.1, left) is $\exp(-((x-x_0)/\sigma_x)^{10} - ((y-y_0)/\sigma_y)^2)\sin(q(x-x_0))$ with a spread $\sigma_x = 2.75$ and $\sigma_y = 2$, is centered at $(x_0, y_0) = (3.5, 5.5)$ and has energy $\omega = q = 5$. Initially this wavepacket approaches the two rectangular blocks of dielectric material (permittivity $\varepsilon = 5$, permeability $\mu = 1$) from the left.

In Table 26.1 we present the results of an error analysis and arithmetic-operation count. The calculations have been carried out for the same system as in Fig. 26.1. The fourth-order unconditionally stable algorithm $T4S2$ [3,4] with a time step τ was used to compute $\Psi(t,\tau)$. From Table 26.1 we see that the error of T4S2 decreases as τ^4, as it should [13]. For $\tau = 0.001$ the result of T4S2 and of the one-step algorithm are about the same to working precision.

Depending on the desired accuracy of the T4S2 calculation, the comparison of the N_H's of the one-step and T4S2 algorithm shows that the former can be orders of magnitude more efficient. Also the Yee FDTD algorithm is no match for the one-step algorithm from the point of view of efficiency.

Although we have not yet made serious efforts to optimize the code, typically the one-step algorithm is more than an order of magnitude faster than FDTD algorithms [16]. This roughly matches our expectations based on a count of the number of arithmetic operations for the two methods, taking as input the value of K on the one hand, and the number of time steps and the order of the FDTD algorithms on the other. Our general conclusions are in concert with those drawn on the basis of numerical experiments with the Schrödinger equation [9].

Acknowledgements

K.M. and H.D.R. are grateful to T. Iitaka for drawing our attention to the potential of the Chebyshev method and for illuminating discussions. This work is partially supported by the Dutch 'Stichting Nationale Computer Faciliteiten' (NCF).

References

1. A. Taflove and S.C. Hagness: *Computational Electrodynamics – The Finite-Difference Time-Domain Method* (Artech House, Boston 2000)
2. K.S. Yee: IEEE Transactions on Antennas and Propagation **14**, 302 (1966)
3. J.S. Kole, M.T. Figge and H. De Raedt: Phys. Rev. E **64**, 066705 (2001)
4. J.S. Kole, M.T. Figge and H. De Raedt: submitted to Phys. Rev. E
5. J.S. Kole, M.T. Figge and H. De Raedt: this volume
6. O.P. Gandi: *Advances in Computational Electrodynamics – The Finite-Difference Time-Domain Method*, A. Taflove, Ed. (Artech House, Boston 1998)
7. B. Houshmand, T. Itoh, and M. Piket-May: *Advances in Computational Electrodynamics – The Finite-Difference Time-Domain Method*, A. Taflove, Ed. (Artech House, Boston 1998)
8. H. Tal-Ezer and R. Kosloff: J. Chem. Phys. **81**, 3967 (1984)
9. C. Leforestier, R.H. Bisseling, C. Cerjan, M.D. Feit, R. Friesner, A. Guldberg, A. Hammerich, G. Jolicard, W. Karrlein, H.-D. Meyer, N. Lipkin, O. Roncero, and R. Kosloff: J. Comp. Phys. **94**, 59 (1991)
10. R.N. Silver and H. Röder: Phys. Rev. E **56**, 4822 (1997)
11. T. Iitaka, S. Nomura, H. Hirayama, X. Zhao, Y. Aoyagi, and T. Sugano: Phys. Rev. E **56**, 1222 (1997)
12. Y.L. Loh, S.N. Taraskin, and S.R. Elliot: Phys. Rev. Lett. **84**, 2290 (2000)
13. H. De Raedt: Comp. Phys. Rep. **7**, 1 (1987)
14. J.H. Wilkinson, *The Algebraic Eigenvalue Problem* (Clarendon Press, Oxford 1965)
15. M. Abramowitz and I. Stegun: Handbook of Mathematical Functions (Dover, New York 1964)
16. This conclusion also holds for full three-dimensional calculations: H. De Raedt, K. Michielsen, J.S. Kole, and M.T. Figge, to be published

27 Statistical Mechanics of Classical Distinguishable Particles

R.H. Swendsen[1]

Physics Department, Carnegie Mellon University, Pittsburgh, PA, 15213, USA

Abstract. An alternative definition of the entropy in terms of the logarithm of the probability distribution of thermodynamic variables is proposed, which is consistent with all desired thermodynamic properties. The factor of $1/N!$ connected with Gibbs' Paradox is shown to arise naturally for distinguishable particles.

27.1 Introduction

Since an essential goal of computer simulations is comparison with the results of real experiments, it is important to understand the relationship between classical models of the properties of distinguishable particles (as they are numbered in a computer simulation) and the properties of real, intrinsically quantum systems in the classical limit. It is often asserted in textbooks that the entropies of classical systems of distinguishable and indistinguishable particles differ by a term involving the logarithm of $N!$. If this were true, results obtained from computer simulations would have to be corrected with a corresponding term. In fact, there is no difference in the properties of distinguishable and indistinguishable particles for classical systems, and the usual expression for the entropy of a system of distinguishable particles is incorrect. In this paper, I discuss a new definition of the entropy that agrees with the thermodynamic definition and makes correct predictions.

The usual expression for the entropy is based on the work of Boltzmann [1], with essential contributions by Gibbs [2], which led to a conjecture that the entropy was proportional to the logarithm of a volume in phase space.

One problem with this definition is the well-known Gibbs' Paradox [2], which is closely related to the fact that Boltzmann's entropy is not extensive. The standard correction is to include a factor of $1/N!$, where N is the number of microscopic particles, to restore extensivity. This correction, known as "correct Boltzmann counting," is usually justified by arguing that particles in quantum mechanics are fundamentally indistinguishable. This has the consequence that the entropy of a system of truly distinguishable particles, such as those in computer simulations, would not have such a factor. As shown below, this leads to predictions for the properties of a system of distinguishable particles that are clearly wrong.

The only direct statements I have found in the literature that the factor of $1/N!$ should be included for distinguishable particles occur in papers by van

Kampen [3] and Jaynes [4]. Their arguments are correct, but their approaches are somewhat different than the one I present here.

27.2 The Thermodynamic Postulates

The essential properties of the entropy are (paraphrasing the postulates given by Callen [5]):

1. The values assumed by the extensive parameters of a composite system in the absence of an internal constraint are those that maximize the entropy over the manifold of constrained equilibrium states.
2. The entropy of a composite system is additive over the constituent subsystems.

The first is the defining property of the entropy, and is equivalent to the second law of thermodynamics. The second is almost as important, since it is needed to allow us to define the entropy of a subsystem in a meaningful way. Other properties (a monotonically increasing function of the energy, continuous, differentiable, and extensive) are extremely useful, but not as essential for defining the entropy.

Additivity and extensivity are sometimes confused. An example of the distinction is given by a gas in which some of the gas molecules are adsorbed on the walls of the container. The entropy of such a common system is not extensive, but it is additive.

27.3 Boltzmann's Conjecture

Boltzmann (and Gibbs) conjectured that the entropy of a system of a large number N of particles was given by the logarithm of the accessible volume of phase space Ω_B.

$$S_B = k_B \ln \Omega_B \,, \tag{27.1}$$

defined as including all energies less than some maximum energy E, and all particle positions in a given volume V [1,2].

This assumption leads to the expression for the entropy of a classical ideal gas of distinguishable particles

$$S_B = k_B N \left[\ln(V) + \left(\frac{3}{2}\right) \ln\left(\frac{E}{N}\right) + X_B \right] , \tag{27.2}$$

where X_B is a constant independent of E, V, and N. Gibbs' Paradox is the observation that when mixing two containers of such gases, the Boltzmann entropy does not behave consistently. If the system is doubled in size, S_B becomes $2S_B + k_B N ln 2$.

It is usually claimed that we can "correct" (27.1) – if the particles are indistinguishable – by dividing Ω_B by $N!$, giving an expression for the entropy that behaves reasonably.

If this argument were correct, it would imply that the entropy for distinguishable particles should be (27.1). One way to see that this is not true is to consider of two ideal gases that can exchange particles and energy. The total Boltzmann entropy from (27.2) is then

$$S_B = k_B N_1 \left[\ln(V_1) + \left(\frac{3}{2}\right) \ln\left(\frac{E_1}{N_1}\right) + X_B \right] \\ + k_B N_2 \left[\ln(V_2) + \left(\frac{3}{2}\right) \ln\left(\frac{E_2}{N_2}\right) + X_B \right] , \tag{27.3}$$

with $E = E_1 + E_2$, etc. Setting the partial derivative with respect to E_1 equal to zero gives the usual equilibrium condition $E_1/N_1 = E_2/N_2$ as expected.

However, the partial derivative with respect to N_1 leads to

$$\frac{\partial S_B}{\partial N_1} = k_B \ln\left(\frac{V_1}{V_2}\right) . \tag{27.4}$$

Since the right hand side a constant, the Boltzmann entropy has its maximum either at $N_1 = N$ or $N_1 = 0$, giving the bizarre prediction that all particles should go to the larger of the two subvolumes! The Boltzmann entropy is clearly incorrect.

27.4 An Alternative Definition of the Entropy

If we define the entropy in statistical mechanics as the logarithm of the probability distribution for the experimentally accessible observables in a composite system, the conflicts are resolved, and the validity of the second law of thermodynamics is immediately apparent.

This definition only requires the validity of the first property of the entropy. Other properties (additivity, monotonicity, continuity and differentiability, and extensivity) are not assumed, but can be derived for those systems to which they are applicable.

To see the consequences of this definition, first consider a classical ideal gas. The positions and momenta are assumed to be independent, so that we can analyze them separately.

Confining particles to a volume V and assuming a uniform probability density, we find that the distribution of particles between two subvolumes V_1 and V_2 is given by the binomial distribution

$$W(N_1, N_2) = \frac{N!}{N_1! N_2!} \left(\frac{V_1}{V}\right)^{N_1} \left(\frac{V_2}{V}\right)^{N_2} \tag{27.5}$$

with the added constraint that $N_1 + N_2 = N$. This leads to the correct equilibrium solution $\langle N_1 \rangle = N V_1 / V$.

The logarithm of the probability distribution in (27.5) can be written in the interesting form

$$\ln W(N_1, N_2) = \ln \Omega_q(V_1, N_1) + \ln \Omega_q(V_2, N_2) - \ln \Omega_q(V, N)\,, \qquad (27.6)$$

where $\Omega_q(V, N) = V^N/N!$, and the subscript q refers to the positions. The last term in (27.6) is independent of N_1 or V_1 and is not needed to find the maximum of $lnW(N_1, N - N_1)$. This suggests that we identify a "configurational entropy" as $S_q = k_B \ln \Omega_q(V, N)$. The sum of the configurational entropies for the two subsystems is then a maximum at the correct equilibrium value given by $\langle N_1 \rangle = NV_1/V$, in contrast with incorrect prediction of the Boltzmann entropy.

The derivation of the configurational entropy given above solves the problem of the origin of the factor of $1/N!$. Extending the derivation to the full entropy to include integrations over the momenta and going beyond the ideal gas to a general Hamiltonian $H(\{p_i, q_i\})$ is straightforward. Assuming that the total energy is specified exactly, the full expression for the entropy can be shown to be $S = k_B \ln(\Omega)$, where

$$\Omega(E, V, N) = \frac{1}{h^{3N}} \frac{1}{N!} \int d\vec{q}^{\,N} \int d\vec{p}^{\,N} \delta\left(E - H\left(\{p_i, q_i\}\right)\right) . \qquad (27.7)$$

Here, I have included the usual factor of $1/h^{3N}$ to make contact with the classical limit of quantum statistics.

The entropy obtained from (27.7) is valid for all classical systems of distinguishable particles. Although I do not have space to demonstrate it in this paper, the same equation is applicable to classical systems of indistinguishable particles. Classical particles do not care whether we know their names.

In summary, despite the importance of Boltzmann's conjecture, it contains a fundamental flaw that has practical consequences for computer simulations. The definition of the entropy I have proposed corrects this flaw and leads to predictions that are consistent with the thermodynamic postulates.

References

1. L. Boltzmann: Sitzb. Akad. Wiss. Wien, **76**, 373 (1877), *Wissenschaftliche Abhandlungen von Ludwig Boltzmann*, Vol. II (Chelsea, New York 1968) pp. 164–223
2. J.W. Gibbs: *Elementary Principles in Statistical Mechanics* (Yale University Press, New Haven 1902); reprinted by (Dover, New York 1960)
3. N.G. van Kampen: "The Gibbs Paradox," in *Essays in Theoretical Physics*. Ed. by W.E. Parry (Pergamon, Oxford 1984) pp. 303–312
4. E.T. Jaynes: "The Gibbs Paradox," in *Maximum-Entropy and Bayesian Methods*. Ed. by G. Erickson, P. Neudorfer, and C.R. Smith (Kluwer, Dordrecht 1992)
5. H.B. Callen: *Thermodynamics and an Introduction to Thermostatistics* (Wiley, New York 1985)

List of Contributors

SPRINGER PROCEEDINGS IN PHYSICS

50 **Magnetic Properties of Low-Dimensional Systems II**
New Developments
Editors: L.M. Falicov, F. Mejía-Lira, and J.L. Morán-López

51 **The Physics and Chemistry of Organic Superconductors**
Editors: G. Saito and S. Kagoshima

52 **Dynamics and Patterns in Complex Fluids**
New Aspects of the Physics–Chemistry Interface
Editors: A. Onuki and K. Kawasaki

53 **Computer Simulation Studies in Condensed-Matter Physics III**
Editors: D.P. Landau, K.K. Mon, and H.-B. Schüttler

54 **Polycrystalline Semiconducturs II**
Editors: J.H. Werner and H.P. Strunk

55 **Nonlinear Dynamics and Quantum Phenomena in Optical Systems**
Editors: R. Vilaseca and R. Corbalán

56 **Amorphous and Crystalline Silicon Carbide III, and Ohter Group IV–IV Materials**
Editors: G.L. Harris, M.G. Spencer, and C.Y. Yang

57 **Evolutionaly Trends in the Physical Sciences**
Editors: M. Suzuki and R. Kubo

58 **New Trends in Nuclear Collective Dynamics**
Editors: Y. Abe, H. Horiuchi, and K. Matsuyanagi

59 **Exotic Atoms in Condensed Matter**
Editors: G. Benedek and H. Schneuwly

60 **The Physics and Chemistry of Oxide Superconductors**
Editors: Y. Iye and H. Yasuoka

61 **Surface X-Ray and Neutron Scattering**
Editors: H. Zabel and I.K. Robinson

62 **Surface Science**
Lectures on Basic Concepts and Applications
Editors: F.A. Ponce and M. Cardona

63 **Coherent Raman Spectroscopy**
Recent Advances
Editors: G. Marowsky and V.V. Smirnov

64 **Superconducting Devices and Their Applications**
Editors: H. Koch and H. Lübbing

65 **Present and Future of High-Energy Physics**
Editors: K.-I. Aoki and M. Kobayashi

66 **The Structure and Conformation of Amphiphilic Membranes**
Editors: R. Lipowsky, D. Richter, and K. Kremer

67 **Nonlinearity with Disorder**
Editors: F. Abdullaev, A.R. Bishop, and S. Pnevmatikos

68 **Time-Resolved Vibrational Spectroscopy V**
Editor: H. Takahashi

69 **Evolution of Dynamical Structures in Complex Systems**
Editors: R. Friedrich and A. Wunderlin

70 **Computational Approaches in Condensed-Matter Physics**
Editors: S. Miyashita, M. Imada, and H. Takayama

71 **Amorphous and Crystalline Silicon Carbide IV**
Editors: C.Y. Yang, M.M. Rahman, and G.L. Harris

72 **Computer Simulation Studies in Condensed-Matter Physics IV**
Editors: D.P. Landau, K.K. Mon, and H.-B. Schüttler

73 **Surface Science**
Principles and Applications
Editors: R.F. Howe, R.N: Lamb, and K. Wandelt

74 **Time-Resolved Vibrational Spectroscopy VI**
Editors: A. Lau, F. Siebert, and W. Werncke

75 **Computer Simulation Studies in Condensed-Matter Physics V**
Editors: D.P. Landau, K.K. Mon, and H.-B. Schüttler

76 **Computer Simulation Studies in Condensed-Matter Physics VI**
Editors: D.P. Landau, K.K. Mon, and H.-B. Schüttler

77 **Quantum Optics VI**
Editors: D.F. Walls and J.D. Harvey

78 **Computer Simulation Studies in Condensed-Matter Physics VII**
Editors: D.P. Landau, K.K. Mon, and H.-B. Schüttler

MIX
Papier aus verantwortungsvollen Quellen
Paper from responsible sources
FSC® C105338

If you have any concerns about our products,
you can contact us on
ProductSafety@springernature.com

In case Publisher is established outside the EU,
the EU authorized representative is:
Springer Nature Customer Service Center GmbH
Europaplatz 3, 69115 Heidelberg, Germany

Printed by Libri Plureos GmbH
in Hamburg, Germany